Digital Circuit SIMULATION Using Excel

Digital Circuit SIMULATION Using Excel

ANTHONY MAZZURCO

Columbus, Ohio

The views and opinions expressed in this book are solely those of the author and do not reflect the views or opinions of Gatekeeper Press. Gatekeeper Press is not to be held responsible for and expressly disclaims responsibility of the content herein.

Digital Circuit Simulation Using Excel

Published by Gatekeeper Press
2167 Stringtown Rd, Suite 109
Columbus, OH 43123-2989
www.GatekeeperPress.com

Copyright © 2020 by Anthony Mazzurco

All rights reserved. Neither this book, nor any parts within it may be sold or reproduced in any form or by any electronic or mechanical means, including information storage and retrieval systems without permission in writing from the author. The only exception is by a reviewer, who may quote short excerpts in a review.

ISBN (paperback): 9781662902499
eISBN: 9781662902505

*To Sandy,
my wife and best friend,
in gratitude for her unflagging support and encouragement.*

Contents

Introduction 1

Chapter 1: The Representation of Logical Operators 5

    Basic Circuit Functions 5

    The Inverter 5

    The Two Input AND Gate 6

    The Two Input OR Gate 7

    The Two Input NAND Gate 8

    The Two Input EXCLUSIVE-OR Gate 11

    Remaining Operators 12

        The Two Input NOR Gate 12

        The Two Input EXCLUSIVE-NOR Gate 12

Chapter 2: Simulation of the Basic Logical Operators 15

    INVERTER 15

    AND Gate 17

    NAND Gate 21

    OR Gate 23

    EXCLUSIVE-OR Gate 25

    NOR Gate 29

    EXCLUSIVE-NOR Gate 32

Chapter 3: Simulation of Combinatorial Logic Circuits 35

Chapter 4: Memory Devices 51
R-S Latch 51
D Flip Flop 61
J-K Flip Flop 81

Chapter 5: Asynchronous Design 91

Chapter 6: Synchronous Design 111

Chapter 7: Considerations for Propagation Delay and Timing Analysis 133
NAND Gate 133
Combinatorial Logic 139
Flip Flops 150
Asynchronous Counter 153
Synchronous Counter 164

Chapter 8: Variable Delay Analysis 179

Chapter 9: Excel Implementation of Standard 7400 Devices 195
Tri-State Devices 195
Decoders 208
One Shot (Monostable Multivibrator) 213

Bibliography 227
Index 229

Introduction

Simulation plays an important role in the process of circuit development. There are obvious cost benefits that derive from being able to perform simulation, verification, and troubleshooting before hardware is developed.

There are software packages available to simulate both analog and digital circuitry. Once engineers become proficient in them, subtle design problems can be found and addressed before prototypes are produced.

The approach presented here is to introduce a method that focuses on a straightforward method for simulating digital circuitry that requires no specialized training. The simulation uses Microsoft's Excel.

Since Excel is found on many computers, most users already have at least a basic understanding of how it works. So the learning process is minimal.

The techniques developed in this book can be useful to practicing engineers who want to perform simulations of their design to find logical and timing errors before any hardware is built. It would also be of value in a classroom setting to aid in teaching the principles of digital circuit design by introducing students to the subtleties of combinatorial logic design and timing analysis. It allows digital circuits to be tested over a range of component timing specifications that will expose limitations in the design.

The simulation tools are developed through the following topics:

In the first chapter, a general review of the common SSI digital circuit elements is covered. This includes the basic gates such as NAND, AND, OR, etc. This discussion is very generic, focusing on the logical operation of these devices. It covers truth tables, circuit symbols, and their representation in logical expressions.

The second chapter introduces how the basic circuit elements are represented in Excel. Spreadsheets are developed to verify the truth tables associated with each device.

Chapter 3 shows how the basic gates are combined in a spreadsheet to simulate a circuit. The resulting simulation verifies the static combinatorial input/output relationship. This chapter also shows how a library of functions simulating the basic gates is developed. There is a brief introduction into how the spreadsheets can be used to simulate inputs and outputs that are time-varying (i.e. dynamic) signals.

Chapter 4 shows how memory components (latches and flip flops) are represented in spreadsheets. The basis for simulating sequential circuity is developed.

Chapter 5 uses the design for an asynchronous circuit to show how it is simulated in a spreadsheet. This case study demonstrates how propagation delays are modeled.

Another design is studied in Chapter 6. Here, a synchronous up/down counter is developed and simulated. Timing perturbations to the inputs are introduced and the simulation is analyzed to see the behavior of the circuit under these conditions.

Chapter 7 shows how propagation delays through circuit components can cause switching hazards and how the simulation can be used to discover them. Each of the designs from the previous two chapters is revisited to expose and find the root causes for some of the unexpected results that can occur once we introduce propagation delays.

Chapter 8 takes the timing analysis one step further to show how the spreadsheet can rigorously test circuits by simulating combinations of propagation delays (from typical to worst case) through the various signal paths.

In the last chapter, the simulation models for a few additional devices are developed to show how the user can add more components to the library. The text focuses on the 74LS family of devices. However, any of the standard logic families (74F, 74HC, etc.) can be added with only a minor parameter modification to accommodate the propagation delay.

There are limitations to the Excel simulation technique. For example, it does not check for loading errors like fan-out. Also, analog-related problems are not addressed. These would include possible inductive coupling between signals, transmission line effects, and problems that can arise from the circuit layout.

Additional background and supporting reference information on some of the topics covered are cited at the end of each chapter. The referenced documents are those listed in the bibliography.

CHAPTER

1

The Representation of Logical Operators

Basic Circuit Functions

Digital circuitry design is based on implementing logical statements comprised of variables associated by "logical operators" or "logical connectives." The variables are binary and therefore take on a value of either 0 or 1. The operators define a function performed on independent binary variables to produce a result which is also binary.

The following is a review of some of the basic operators. We will show their common circuit representations, logic equation symbols, and truth tables.

The Inverter

This is a single variable operator which converts a 1 to a 0 and a 0 to a 1. That is, it provides an output whose state is the logical "inversion" of its input's state. Below, the circuit symbol for the device is shown where the logical signal flow is left to right.

In logical equations, the inversion operation is shown either with a tilde in front of or a bar above the independent variable (i.e. the input). So if the independent variable were X and the dependent (output) variable were Y, the inversion of X would be shown as either

$$\sim X = Y$$

or

$$\overline{X} = Y$$

The truth table defining the input/output relationship is straightforward and is given as:

INPUT (X)	OUTPUT ($\overline{X}$)
0	1
1	0

Since a single independent variable only has two states, there are only two lines in the truth table.

The Two Input AND Gate

This basic two-input AND gate operation says that when both inputs are 1, the output is a 1. Otherwise, the output is a 0. Its circuit symbol is:

Chapter 1: *The Representation of Logical Operators* | 7

The AND operation is usually represented with a dot between the variables being "AND-ed" together. So if the input variables are X and Y and the output is Z, the AND operation is represented as:

$$X \cdot Y = Z$$

Sometimes the dot is omitted and the input variables of the AND operation are just concatenated as:

$$XY = Z$$

Its truth table is:

INPUT 1 (X)	INPUT 2 (Y)	OUTPUT (X·Y) = Z
0	0	0
0	1	0
1	0	0
1	1	1

Since there are two independent, two state input variables, there are now a total of four combinations in the truth table. In general, the total number of combinations that can be realized for "n" number of two state variables is 2^n. So here $2^2 = 4$.

If the AND gate has more than two inputs there will be additional rows in the truth table but the output will only be a 1 at the single instance where all of the inputs are simultaneously 1.

The Two Input OR Gate

The two input OR gate produces an output of 1 whenever either (or both) of its two inputs is a 1.

INPUT 1 ─────┐
             ├──D──── OUTPUT
INPUT 2 ─────┘

The OR operation is represented with a plus sign. With input variables X and Y and output Z, the OR function is expressed as:

$$X + Y = Z$$

The truth table for the OR operation is:

INPUT 1 (X)	INPUT 2 (Y)	OUTPUT (X + Y)
0	0	0
0	1	1
1	0	1
1	1	1

If there were more than two inputs, the number of lines in the table would increase to a total number of 2^n as described above, where "n" is the number of inputs. But regardless of how many inputs it has, the only place where there would be a 0 at the output of the OR gate would be where all of the input variables are zero.

The Two Input NAND Gate

This is a very common operator found in practice. This is simply the AND function with an inverted output. The symbol is the same as the AND gate, but with a bubble at the output which is a shorthand representation to show that there is an inversion at this point in the gate. The bubble is not used as a standalone symbol.

Its formula representation is an inversion over the AND-ing of the two inputs. For inputs X and Y and output Z we have

$$\sim(X \cdot Y) = Z$$

The truth table is

INPUT 1 (X)	INPUT 2 (Y)	OUTPUT ~(X · Y) = Z
0	0	1
0	1	1
1	0	1
1	1	0

Clearly, if there are more inputs, the only time that the output would be zero is when all of the inputs are one.

This is a very versatile function. It can be used to simulate any of the other logical operations we have already mentioned. For example, if inputs X and Y are connected together so that they can only have the same value, the gate would operate as an inverter. This is verified by looking at the first and last rows of the preceding NAND truth table. Specifically, the table reduces to:

INPUT 1 (X)	INPUT 2 (Y)	OUTPUT ~(X · Y) = Z
0	0	1
1	1	0

In addition, by putting an inverter after the NAND function, we have implemented an AND.

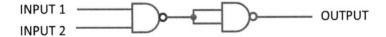

For implementing the OR function, DeMorgan's theorem is used. This theorem states that the inversion of an AND-ing operation between two input signals is logically equivalent to the

OR-ing of each input signal's inversion. Represented as an equation, it states:

$$\sim(A \cdot B) = \sim A + \sim B$$

This can be verified by a truth table showing all four input combinations:

A	B	~A	~B	A · B	~(A · B)	~A + ~B
0	0	1	1	0	1	1
0	1	1	0	0	1	1
1	0	0	1	0	1	1
1	1	0	0	1	0	0

The last two columns have the same values in all rows, demonstrating the validity of the theorem.

Functionally, this says that the NAND function is equivalent to the "Invert-OR" function where an "Invert-OR" is an OR gate with its inputs inverted before the OR is applied. The circuit representation is:

INPUT 1 ────┐
            ├─▶ OUTPUT
INPUT 2 ────┘

So, in order to implement an OR function with a NAND (or equivalently an Invert-OR) we need to invert the inputs to the gate as shown:

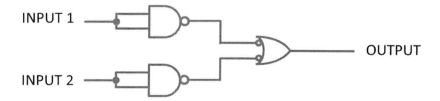

The Two Input EXCLUSIVE-OR Gate

This function is similar to the OR gate with the exception that it "excludes" the case of a 1 at the output when both outputs are 1.

The schematic representation for the gate is:

```
INPUT 1 ────┐
            )D─── OUTPUT
INPUT 2 ────┘
```

The logical operation is represented with the OR plus sign enclosed in a circle. For inputs X and Y with output Z it is:

$$X \oplus Y = Z$$

Its truth table is:

INPUT 1 (X)	INPUT 2 (Y)	OUTPUT (X $\oplus$ Y) = Z
0	0	0
0	1	1
1	0	1
1	1	0

There are other ways looking at this function. One way, when there are just two inputs, is as a two bit comparator. It compares the logical state of its two inputs. When they are different, it outputs a 1. When they are the same, it outputs a 0. Another more general interpretation is that the Exclusive-Or function acts as a parity check over the number of 1s it sees on its inputs. If there are an odd number of 1s, it outputs a 1. An even number of 1s gives a zero. We will come back to this in the next chapter where we will discuss the Exclusive-Or function with more than two inputs. **(How can the Exclusive-Or function be implemented with only NAND gates?)**

Remaining Operators

All of the previously described functions comprise the common operators that are most often used in practice. Two more should be mentioned. They are the NOR and EXCLUSIVE-NOR gates. Their logic symbols, formula representation, and truth tables should be clear from the previous discussion.

The Two Input NOR Gate

Symbol

```
INPUT 1 (X) ─────┐
                 )D─── OUTPUT (Z)
INPUT 2 (Y) ─────┘
```

Formula Representation

$$\sim(X + Y) = Z$$

Truth Table

INPUT 1 (X)	INPUT 2 (Y)	OUTPUT ~(X + Y)=Z
0	0	1
0	1	0
1	0	0
1	1	0

(Can the NOR gate be used the same way as the NAND gate to simulate the other basic logical functions?)

The Two Input EXCLUSIVE-NOR Gate

Symbol

```
INPUT 1 (X) ─────┐
                 )D─── OUTPUT (Z)
INPUT 2 (Y) ─────┘
```

Formula Representation

$$\sim(X \oplus Y) = Z$$

Truth Table

INPUT 1 (X)	INPUT 2 (Y)	OUTPUT $\sim(X \oplus Y)=Z$
0	0	1
0	1	0
1	0	0
1	1	1

There are many excellent sources of information covering gates and their basic logical operations. Further descriptions for the gates discussed in this chapter along with their symbols and truth tables can be found in References [1], [2], [3], [4], [6], [8], and [9].

The use of the NAND and NOR as universal gates is discussed in References [5], [6], [8], and [9].

Axioms and theorems governing logical operations are covered in References [3], [4], and [5]. These include association, commutation, and distribution along with DeMorgan's Theorem.

CHAPTER

2

Simulation of the Basic Logical Operators

The binary logical operations that were reviewed in the previous chapter are simulated in Excel.

INVERTER

First consider the INVERTER. Recall that the INVERTER takes a 0 or 1 as its input and inverts it to a 1 or 0, respectively, at its output.

Consider the following formula view of a spreadsheet (Figure 2-1).

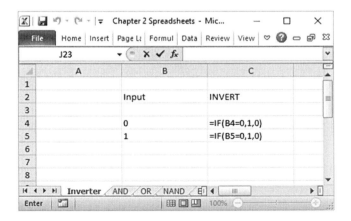

Figure 2-1: Formula View of the INVERTER Simulated in a Spreadsheet

16 | Digital Circuit Simulation Using Excel

Two columns of the spreadsheet are used. In Row 2, each column's function is labeled. Column B contains the input to the INVERTER. Column C gives the resulting output which we label as "INVERT".

There are only two states possible for the input. Naturally they are 0 and 1 and entered in Rows 4 and 5.

The inversion operation is entered manually in Row 4. The operation is implemented by using the IF function in Cell C4. The operation takes its input from Column B. In Row 4, the IF statement says that if the input in B4 is 0 (the first field in the IF statement), then the output is a 1 (the second field in the IF statement), otherwise it is a 0 (the third field in the IF statement).

This takes care of Row 4. For Row 5, we don't need to type in the IF statement again, we just need to copy down C4 into Cell C5.

The data view of the spreadsheet is shown in Figure 2-2.

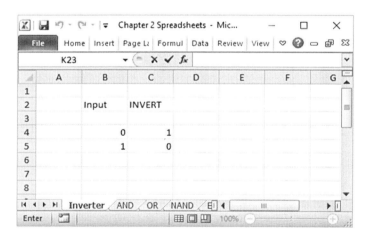

Figure 2-2: Data View of the INVERTER Simulated in a Spreadsheet

We see the familiar truth table for an Inverter.

Note that you can switch between the formula and data views of the spreadsheet by typing [CTRL]~.

Chapter 2: *Simulation of the Basic Logical Operators* | 17

AND Gate

Let us turn next to the AND gate. Here we will introduce a few new aspects to the spreadsheet. The formula view of the AND spreadsheet is shown in Figure 2-3.

Excel provides some logical operators. In this instance, we are using the built-in AND function. This conveniently provides us with the logical operation we are looking for. However, the Excel function does not operate with 0's and 1's but rather with the arguments of TRUE and FALSE. That is not a major problem since we will be able to test our inputs against values of 0 and 1 and make a logical TRUE/FALSE decision from them. Based on whether the conditions are true or not, values of 0 or 1 are assigned.

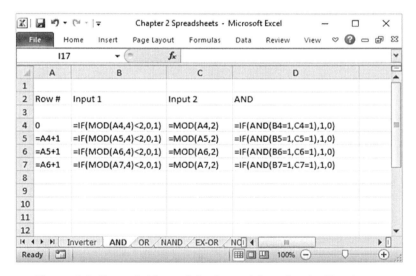

Figure 2-3: Formula View of the Spreadsheet for the Two-Input AND Gate Function

First, notice that we have defined Column A to give a numerical label to the rows starting with row number 0 in Cell A4. After that, the previous row label is incremented (in Row 5 and beyond) for the rest of the truth table. For a 2-input gate having just 4 conditions, there isn't much need to do this. However, when our tables grow, then this will provide a convenient way of automatically generating the input combinations.

The states for the two inputs are generated in Columns B and C starting in Row 4. They are derived from the row number.

Consider Column C which we want to contain the input alternating between 0 and 1 for each row. There are several ways of doing this. We have chosen to simply perform a modulo 2 division on the row label defined in Column A. For even row numbers (including 0) we get a 0 remainder. For odd row numbers we get a 1. So we will see the 0,1,0,1,... sequence generated in each row.

In Column B, representing the 0,0,1,1,... sequence, we first perform a modulo 4 division on the row number. This results in the sequence shown in Table 2-1. Although we are only using 4 rows in the current example, the table is extended in order to show the pattern more clearly.

Row #	(Row #) MOD 4
0	0
1	1
2	2
3	3
4	0
5	1
6	2
7	3
8	0
9	1
10	2

Table 2-1: Applying a Modulo 4 Operation on Row Numbers

The pattern repeats every 4 rows. It can be used to provide a repeating 0,0,1,1 signal for "Input 1" if we set it to a value of 0 if the modulo division gives a result less than 2 while giving a value of 1 otherwise (i.e. 2 or greater).

Finally, in Column D, the logical operation is performed. Remember that the AND gate output is a 1 only if all of the inputs are 1. So looking at Cell D4 we use an IF statement that says, "If both B4 and C4 are a 1, output a 1, otherwise output a 0."

At this point we can copy down Row 4 for Columns B, C, and D to give us our 4 truth table combinations. Notice that Column A must be copied down from Row 5 in order to get the incrementing row numbers automatically. We could have copied down from Row 4 if we put a -1 in Cell A3 and entered

$$=A3+1$$

in Cell A4. We will do this in the next example. First let us just check the data view of the 2-input AND gate spreadsheet as shown in Figure 2-4. It is clear that it correctly performs the AND function.

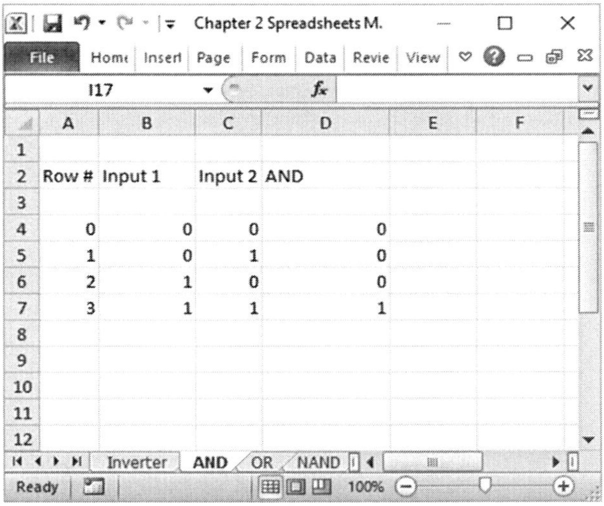

Figure 2-4: Data View of the Spreadsheet for the Two-Input AND Gate Function

20 | Digital Circuit Simulation Using Excel

At this point, let us expand the AND function to a three-input gate. It is very similar to what we have seen for the two-input gate. We do, however, need to add another column for the third input. The formula view of the spreadsheet is shown in Figure 2-5.

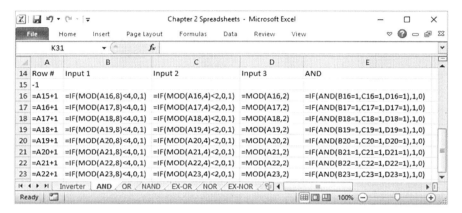

Figure 2-5: Formula View of the Spreadsheet for the Three-Input AND Gate Function

First, notice in Column A that a -1 was entered in the cell above the start of the truth table to allow us to copy down all cells from Row 16 which is the actual start of the table.

Columns C and D have the same formulas as the input columns for the two input AND gate. Column B was added to represent the third input in the truth table. It is patterned after Column C but the parameters are changed to perform a modulo 8 operation with the value set to 0 if the result is less than 4 and 1 if the result is 4 or greater.

Finally, the logical formula for the AND function (Cell E16) was changed to include the extra input. The logical formula is the same as the two-input gate in the sense that all inputs must be 1 in order for the output to be a 1.

The data view of the truth table is shown in Figure 2-6.

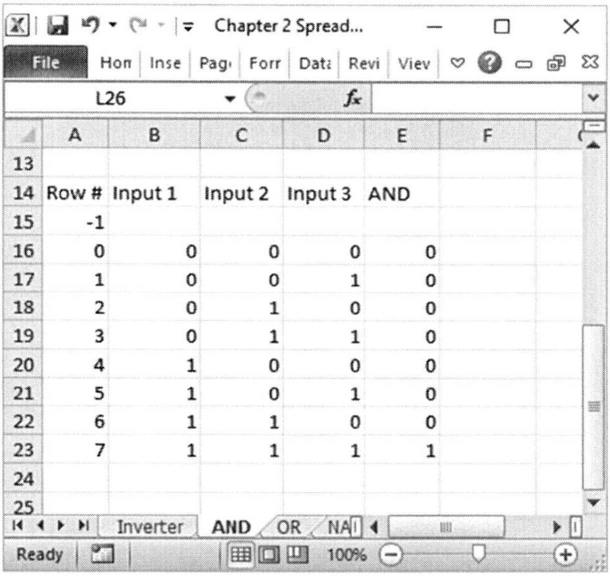

	A	B	C	D	E
13					
14	Row #	Input 1	Input 2	Input 3	AND
15	-1				
16	0	0	0	0	0
17	1	0	0	1	0
18	2	0	1	0	0
19	3	0	1	1	0
20	4	1	0	0	0
21	5	1	0	1	0
22	6	1	1	0	0
23	7	1	1	1	1
24					
25					

Figure 2-6: Data View of the Spreadsheet for the Three-Input AND Gate Function

NAND Gate

The NAND gate spreadsheet description is a simple adaptation of the AND gate. For a given set of inputs, the NAND gate's output is the inverse of the AND. To implement this, the IF/THEN/ELSE statements only need to have the THEN and ELSE actions interchanged. For example, the decision of the two-input AND gate would say:

IF(AND(INPUT1=1, INPUT2=1),1,0)

To create the NAND decision, this would be modified to:

IF(AND(INPUT1=1, INPUT2=1),0,1)

22 | Digital Circuit Simulation Using Excel

The formula and data views of the two-input NAND spreadsheet look like the following:

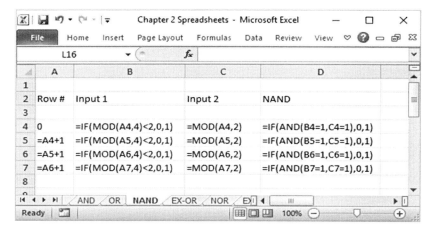

Figure 2-7: Formula View of the Spreadsheet for the Two-Input NAND Gate Function

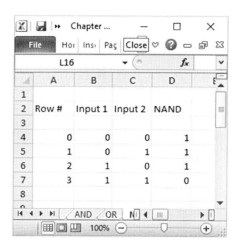

Figure 2-8: Data View of the Spreadsheet for the Two-Input NAND Gate Function

Chapter 2: *Simulation of the Basic Logical Operators* | 23

Extension to more than two inputs follows the same logic as was used for extending the AND gate.

OR Gate

Figure 2-9 shows the spreadsheet formula view for the two-input OR gate.

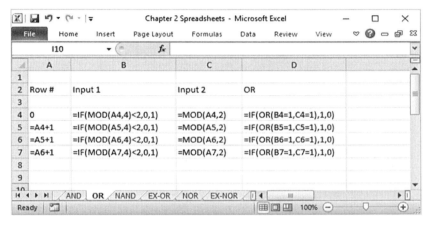

Figure 2-9: Formula View of the Spreadsheet for the Two-Input OR Gate Function

Columns A, B, and C are identical to the same columns used for the two-input AND gate. The difference is in the formula in Column D. Here we test to see if any input (from Columns B and C) is equal to 1. If so, a 1 is output; otherwise a 0 is output. Figure 2-10 shows the data view of the spreadsheet displaying the familiar OR gate truth table.

	A	B	C	D	E
1					
2	Row	Input 1	Input 2	OR	
3					
4	0	0	0	0	
5	1	0	1	1	
6	2	1	0	1	
7	3	1	1	1	
8					
9					

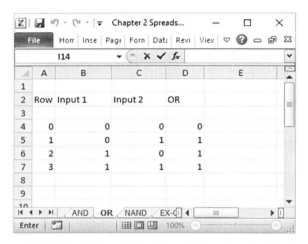

Figure 2-10: Data View of the Spreadsheet for the Two-Input OR Gate Function

The derivation for the three-input OR gate truth table from the two-input gate is analogous to the three-input AND derivation. Figure 2-11 shows the formula view of the spreadsheet while Figure 2-12 gives the data view.

	A	B	C	D	E
13					
14	Row #	Input 1	Input 2	Input 3	OR
15	-1				
16	=A15+1	=IF(MOD(A16,8)<4,0,1)	=IF(MOD(A16,4)<2,0,1)	=MOD(A16,2)	=IF(OR(B16=1,C16=1,D16=1),1,0)
17	=A16+1	=IF(MOD(A17,8)<4,0,1)	=IF(MOD(A17,4)<2,0,1)	=MOD(A17,2)	=IF(OR(B17=1,C17=1,D17=1),1,0)
18	=A17+1	=IF(MOD(A18,8)<4,0,1)	=IF(MOD(A18,4)<2,0,1)	=MOD(A18,2)	=IF(OR(B18=1,C18=1,D18=1),1,0)
19	=A18+1	=IF(MOD(A19,8)<4,0,1)	=IF(MOD(A19,4)<2,0,1)	=MOD(A19,2)	=IF(OR(B19=1,C19=1,D19=1),1,0)
20	=A19+1	=IF(MOD(A20,8)<4,0,1)	=IF(MOD(A20,4)<2,0,1)	=MOD(A20,2)	=IF(OR(B20=1,C20=1,D20=1),1,0)
21	=A20+1	=IF(MOD(A21,8)<4,0,1)	=IF(MOD(A21,4)<2,0,1)	=MOD(A21,2)	=IF(OR(B21=1,C21=1,D21=1),1,0)
22	=A21+1	=IF(MOD(A22,8)<4,0,1)	=IF(MOD(A22,4)<2,0,1)	=MOD(A22,2)	=IF(OR(B22=1,C22=1,D22=1),1,0)
23	=A22+1	=IF(MOD(A23,8)<4,0,1)	=IF(MOD(A23,4)<2,0,1)	=MOD(A23,2)	=IF(OR(B23=1,C23=1,D23=1),1,0)
24					

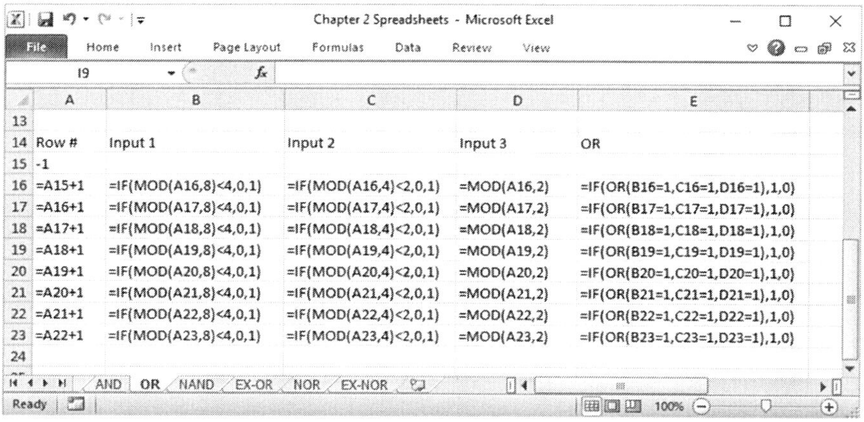

Figure 2-11: Formula View of the Spreadsheet for the Three-Input OR Gate Function

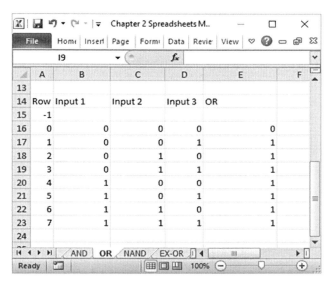

Figure 2-12: Data View of the Spreadsheet for the Three-Input OR Gate Function

EXCLUSIVE-OR Gate

As we have seen in the previous examples, the truth table spreadsheets were derived from applications of the logical functions that are built into Excel. For the AND gate, we used the AND function in the IF/THEN/ELSE statement to define the gate's output. The OR gate uses the OR function in the same way. A complication arises for the EXCLUSIVE-OR (XOR) gate since some of the earlier versions of Excel do not support an XOR function. This does not pose a problem. As noted in the previous chapter, one way of looking at the 2-input Exclusive-Or function is as a comparator between the two inputs. When both inputs are the same, the output is zero. When both inputs are different, the output is one. So the function can be implemented with the following generalized "IF" statement:

$$=IF\ ((INPUT1=INPUT2),0,1)$$

26 | Digital Circuit Simulation Using Excel

This is seen implemented in the formula view given in Figure 2-13.

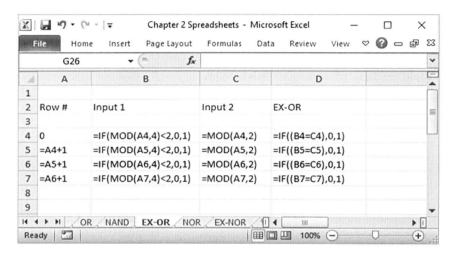

Figure 2-13: Formula View of the Spreadsheet for the Two-Input Exclusive-OR Gate Function

The resulting truth table is given in Figure 2-14.

Row #	Input 1	Input 2	EX-OR
0	0	0	0
1	0	1	1
2	1	0	1
3	1	1	0

Figure 2-14: Data View of the Spreadsheet for the Two-Input Exclusive-OR Gate Function

Considerations for the three-input gate for the Exclusive-Or function rely on the interpretation of the XOR being a parity check function, as mentioned in the previous chapter. That is, the output is a 1 whenever an odd number of the inputs equals 1 (clearly, this interpretation also works for the two-input gate). From this perspective, a truth table can be generated. But let us look at it from a slightly different approach. Let us apply the Associative rule for logical expressions. In the case of the Exclusive-Or function, the Associative rule states:

$$A \oplus B \oplus C = (A \oplus B) \oplus C = A \oplus (B \oplus C)$$

We will simply let the function be defined by taking an Exclusive-Or between the second and third inputs and a second Exclusive-Or between that result and the first input. This is the implementation of

$$A \oplus B \oplus C = A \oplus (B \oplus C)$$

As a schematic, it looks like Figure 2-15.

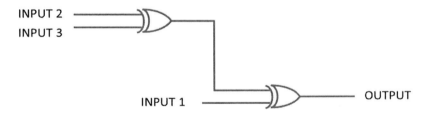

Figure 2-15: Logical Implementation of a Three-Input Exclusive-OR

28 | Digital Circuit Simulation Using Excel

The formula view of the spreadsheet truth table is shown in Figure 2-16.

	A	B	C	D	E	F
13						
14	Row #	Input 1	Input 2	Input 3	EX-OR (COLUMNS C & D)	EX-OR (COLUMNS B & E)
15	-1					
16	=A15+1	=IF(MOD(A16,8)<4,0,1)	=IF(MOD(A16,4)<2,0,1)	=MOD(A16,2)	=IF((C16=D16),0,1)	=IF((B16=E16),0,1)
17	=A16+1	=IF(MOD(A17,8)<4,0,1)	=IF(MOD(A17,4)<2,0,1)	=MOD(A17,2)	=IF((C17=D17),0,1)	=IF((B17=E17),0,1)
18	=A17+1	=IF(MOD(A18,8)<4,0,1)	=IF(MOD(A18,4)<2,0,1)	=MOD(A18,2)	=IF((C18=D18),0,1)	=IF((B18=E18),0,1)
19	=A18+1	=IF(MOD(A19,8)<4,0,1)	=IF(MOD(A19,4)<2,0,1)	=MOD(A19,2)	=IF((C19=D19),0,1)	=IF((B19=E19),0,1)
20	=A19+1	=IF(MOD(A20,8)<4,0,1)	=IF(MOD(A20,4)<2,0,1)	=MOD(A20,2)	=IF((C20=D20),0,1)	=IF((B20=E20),0,1)
21	=A20+1	=IF(MOD(A21,8)<4,0,1)	=IF(MOD(A21,4)<2,0,1)	=MOD(A21,2)	=IF((C21=D21),0,1)	=IF((B21=E21),0,1)
22	=A21+1	=IF(MOD(A22,8)<4,0,1)	=IF(MOD(A22,4)<2,0,1)	=MOD(A22,2)	=IF((C22=D22),0,1)	=IF((B22=E22),0,1)
23	=A22+1	=IF(MOD(A23,8)<4,0,1)	=IF(MOD(A23,4)<2,0,1)	=MOD(A23,2)	=IF((C23=D23),0,1)	=IF((B23=E23),0,1)

Figure 2-16: Formula View of the Spreadsheet for the Three-Input Exclusive-Or Gate Function

Columns A through D of the spreadsheet are the same as for the other three-input gates already covered.

In Column E, we apply the Exclusive-Or operation on the inputs appearing in Columns C and D. A zero is output if they are the same, and a 1 is output if they are different.

Column F performs the second Exclusive-Or. It operates on the Exclusive-Or operation executed in Column E with the input defined in Column B.

Chapter 2: *Simulation of the Basic Logical Operators* | 29

The data view of the spreadsheet is given in Figure 2-17.

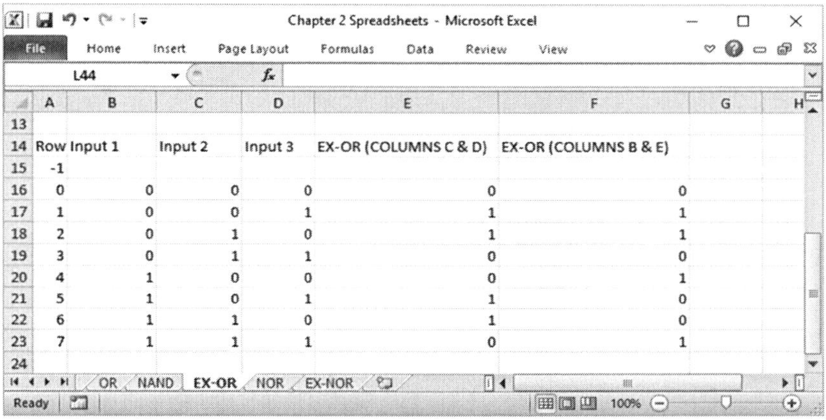

Figure 2-17: Data View of the Spreadsheet for the Three-Input Exclusive-Or Gate Function

The final output is derived in Column F. What we see here is that the function of the gate is consistent with our earlier interpretation of that function being an odd parity check on the number of 1's appearing on its inputs. When there is an odd number of 1's, a 1 is output. When there is an even number of 1's, a 0 is output.

NOR Gate

The spreadsheet implementation of the NOR gate is a straightforward application of the OR gate in much the same way that the NAND gate was derived from the AND gate.

30 | Digital Circuit Simulation Using Excel

The formula view of the spreadsheet for the two-input NOR is given in Figure 2-18.

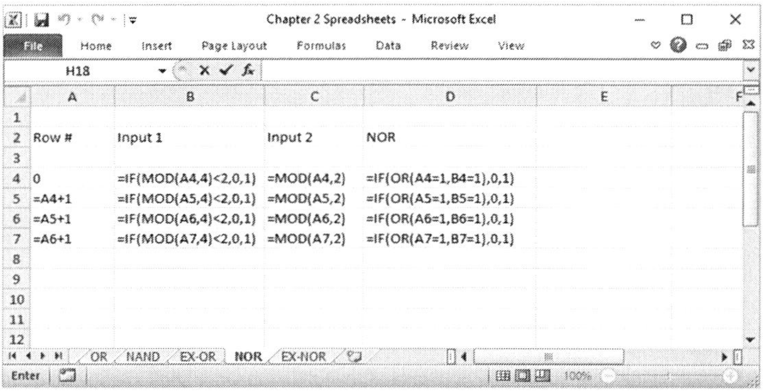

	A	B	C	D
1				
2	Row #	Input 1	Input 2	NOR
3				
4	0	=IF(MOD(A4,4)<2,0,1)	=MOD(A4,2)	=IF(OR(A4=1,B4=1),0,1)
5	=A4+1	=IF(MOD(A5,4)<2,0,1)	=MOD(A5,2)	=IF(OR(A5=1,B5=1),0,1)
6	=A5+1	=IF(MOD(A6,4)<2,0,1)	=MOD(A6,2)	=IF(OR(A6=1,B6=1),0,1)
7	=A6+1	=IF(MOD(A7,4)<2,0,1)	=MOD(A7,2)	=IF(OR(A7=1,B7=1),0,1)

Figure 2-18: Formula View of the Spreadsheet for the Two-Input NOR Gate Function

It is the same as the spreadsheet for the OR gate except that the THEN/ELSE actions taken by the IF statement are reversed. When both or either input is a 1, the output takes on a value of 0.

The data view of the spreadsheet is given in Figure 2-19.

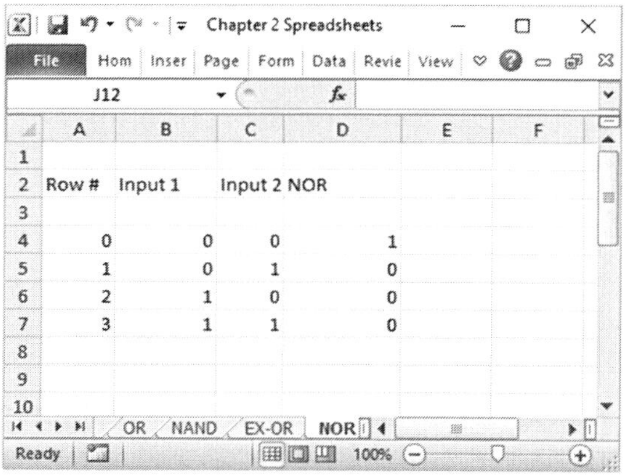

	A	B	C	D
1				
2	Row #	Input 1	Input 2	NOR
3				
4	0	0	0	1
5	1	0	1	0
6	2	1	0	0
7	3	1	1	0

Figure 2-19: Data View of the Spreadsheet for the Two-Input NOR Gate Function

Chapter 2: *Simulation of the Basic Logical Operators* | 31

The formula and data views for the three-input NOR gate are given in Figure 2-20 and Figure 2-21 respectively.

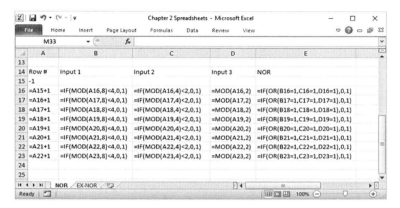

Figure 2-20: Formula View of the Spreadsheet for the Three-Input NOR Gate Function

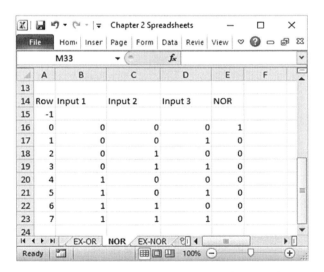

Figure 2-21: Data View of the Spreadsheet for the Three-Input NOR Gate Function

EXCLUSIVE-NOR Gate

We conclude the discussion of the basic gates by covering the Exclusive-NOR gate. Its derivation from the Exclusive-OR gate is completely analogous to the AND to NAND or the OR to NOR situation.

The formula and data views for the two-input gate are given in Figure 2-22 and Figure 2-23 respectively.

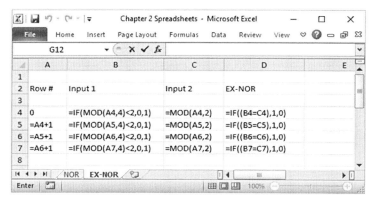

Figure 2-22: Formula View of the Spreadsheet for the Two-Input Exclusive-OR Gate Function

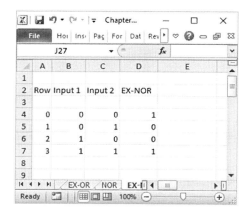

Figure 2-23: Data View of the Spreadsheet for the Two-Input Exclusive-NOR Gate Function

Chapter 2: *Simulation of the Basic Logical Operators* | 33

The formula and data views for the three-input gate are given in Figure 2-24 and Figure 2-25. The only point to note is that, in Figure 2-24, only one Exclusive-OR operation is inverted to get the Exclusive-NOR function. We chose the second operation in Column F between Columns B and E. **(What would happen if we chose to only invert the first operation in Column E (between Columns C and D)?) (What would happen if we inverted both Exclusive-OR operations?)**

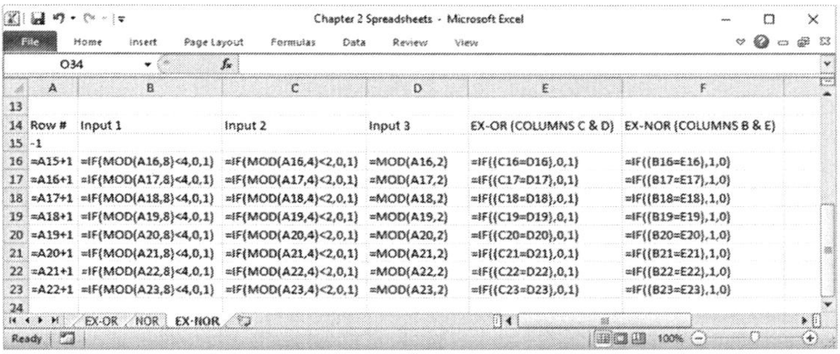

Figure 2-24: Formula View of the Spreadsheet for the Three-Input Exclusive-NOR Gate Function

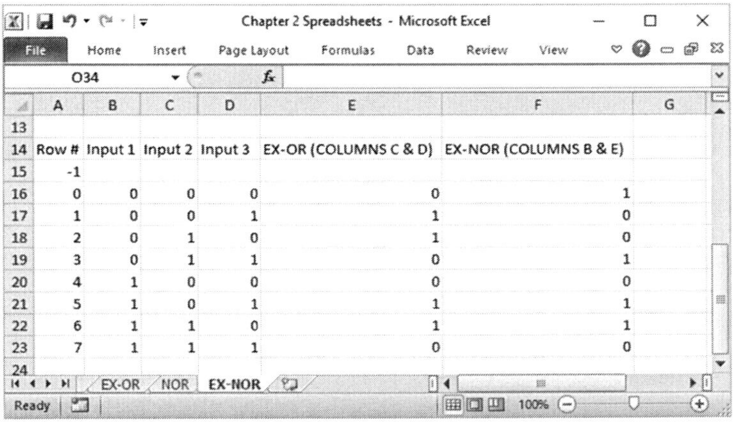

Figure 2-25: Data View of the Spreadsheet for the Three-Input Exclusive-NOR Gate Function

CHAPTER

3

Simulation of Combinatorial Logic Circuits

In the previous chapter, we created spreadsheets in Excel for the basic logic operands. The spreadsheets took defined inputs, performed the logical operation, and generated the output. Truth tables were created by performing the operation over all combinations of static inputs.

In this chapter, the basic circuit elements will be combined using the Excel expressions already developed. Then Visual Basic functions will be developed to simplify the spreadsheet creation.

Consider the simple selector circuit shown in Figure 3-1.

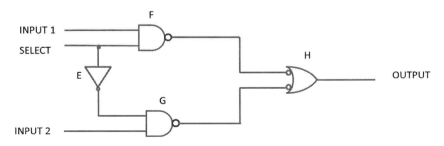

Figure 3-1: Selector Circuit

There are three inputs to the circuit. There are two data lines represented as INPUT 1 and INPUT 2. Either one of them can be selected to be passed through to OUTPUT. The SELECT input determines which one is selected. If SELECT is 1, then INPUT 1 will be presented at OUTPUT. If SELECT is 0, then INPUT 2 will be selected to pass through to OUTPUT.

Using the basic logic functions defined in the previous chapter, a spreadsheet is created to simulate this circuit from those functions. The formula view for the spreadsheet simulation of the circuit is shown in Figure 3-2.

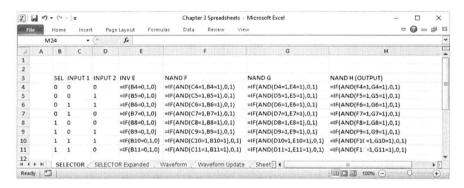

Figure 3-2: Formula View of the Selector Simulation Spreadsheet

First, note how the inputs are defined. For spreadsheet Rows 4 through 7, SELECT is a 0. Therefore, we would expect INPUT 2 to be passed through to the output. Rows 8 to 12 have SELECT as a 1, enabling INPUT 1 to be presented at OUTPUT. For each selection, all combinations of inputs are tested to be certain that the correct input is passed to the output and the value of the unselected input is treated as a "don't care."

Columns E through H represent the output of each of the four gates in Figure 3-1. The Table begins in Row 4. The Boolean expression for the inverter is entered in Column E. It is copied from the INVERTER truth table spreadsheet derived in the last chapter. Columns F, G, and H of Row 4 have the logical expressions for the NAND gates with labels F, G, and H respectively in the circuit drawing. Note that NAND Gate G is shown as an INVERT-OR gate in the schematic (remember DeMorgan's Theorem). This is so that when we view the schematic, the inverting bubbles on the outputs of the two NAND gates logically cancel with the bubbles at the input of the INVERT-OR making it clear that we are OR-ing two AND operations.

All that needs to be done is to ensure that the operations pick up the correct inputs. The INVERTER in Column E picks up its input from Column B (SELECT). NAND Gate F receives Columns B and C, which are the non-inverted value of SELECT and INPUT 1. NAND Gate G inputs Columns D and E, which are INPUT2 and the inverted value of SELECT, respectively. Finally, NAND Gate H inputs Columns F and G, which are the outputs of the first two NAND gates. The output of H is the signal OUTPUT.

To complete the table, Columns E, F, G, and H are copied down from Row 4.

Before looking at the data view of the spreadsheet, we should note that each of its columns actually represents a connection point from Figure 3-1. These connections are either the input or output points of the circuit or interconnecting points between circuit elements. These are the same points available to someone analyzing the physical circuit on a bench using a logic analyzer or oscilloscope.

The data view of the spreadsheet is shown in Figure 3-3.

38 | Digital Circuit Simulation Using Excel

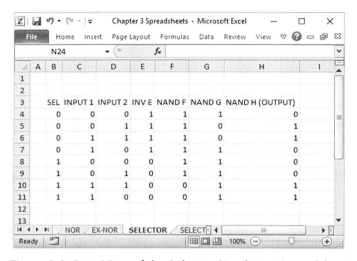

Figure 3-3: Data View of the Selector Simulation Spreadsheet

First, let us look at what happens in Rows 4 to 7. Here, the SELECT input is a 0. So we would expect OUTPUT to reflect INPUT2, regardless of INPUT1. That is the case if we compare Cells D4 to D7 with H4 to H7. When INPUT 2 is a 0 (Rows 4 and 7) OUTPUT is a 0 even though INPUT1 takes on values of 0 (Row 4) and 1 (Row 7). When INPUT2 is a 1 in Rows 5 and 6, OUTPUT is equal to 1 independent of INPUT1 which is a 0 in Row 5 and a 1 in Row 6.

Similarly, when SELECT is a 1 in Rows 8 to 11, OUTPUT is equal to INPUT1 regardless of the state of INPUT2. The verification for this is analogous to the previous situation where SELECT was a 0.

Up to now, the spreadsheets have been used to verify the truth tables defining the circuit. In other words, it is a static verification of the circuit. While useful in that respect, it would also be quite valuable to simulate the operation of the circuit while the inputs are changing dynamically. This will be become especially useful in subsequent chapters when analyzing sequential circuits and their responses to propagation delays and race conditions. Let us digress a moment to see how this will be set up.

Chapter 3: *Simulation of Combinatorial Logic Circuits* | 39

Consider that the input columns represent a progression in time which creates a time-varying waveform at the output. These inputs and outputs would be the waveforms one would see on a logic analyzer. In this situation, each row of the spreadsheet would be considered to represent a fixed period of time which is small in comparison to the rate, or frequency, with which the signals switch but comparable to the expected propagation delays through the circuit elements. For the 74LS family of logic, this would be in the order of about 10 to 20 ns. For this example, let us assign each spreadsheet row a duration of 20 ns. Furthermore, we will assume that the fastest rate for an input signal to switch is 160 ns (or 8 spreadsheet rows). Under these assumptions, the test shown in Figure 3-3 will look like Figure 3-4. Essentially, Figure 3-4 is a 1 to 8 expansion of Figure 3-3.

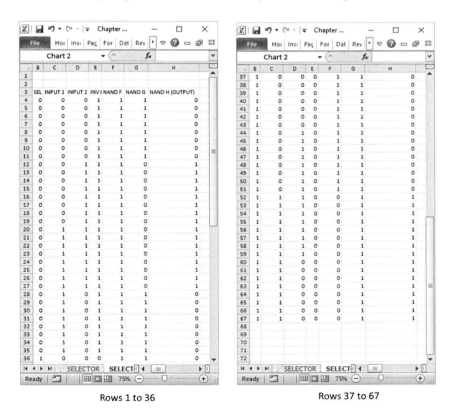

Rows 1 to 36   Rows 37 to 67

Figure 3-4: Expanded Version of Selector Simulation

With the SEL, INPUT1, INPUT2, and OUTPUT columns of Figure 3-3 representing signals that progress in time moving down the table, Figure 3-5 shows timing diagrams for the three inputs and the one output. To create the timing diagrams, each of the desired Columns (B, C, D, and H) was selected and the Excel Insert function for a "3-D Line with Markers" graph was created for each. The datapoints indicated with dots in the graph show the actual values taken from each row of the spreadsheet. We opted to have the datapoints connected with lines to simulate actual signal traces.

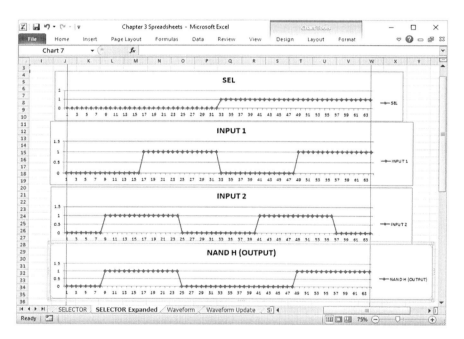

Figure 3-5: Input/Output Timing Diagrams for the Selector Circuit

The figure gives a very good representation of what would appear on a logic analyzer. The only point that isn't addressed is that the outputs change immediately with the input. There is no evidence of propagation delay and, as such, it does not help us analyze potential timing problems. However, after a means of incorporating delays (particularly for sequential circuits) is developed in Chapter 7, the modified spreadsheets and the associated timing diagrams will prove to be useful tools.

At this point, let us return to the discussion of truth tables to introduce another concept that will help simplify their generation.

Going back to one of the spreadsheets, Figure 3-2 for example, we can see the equations for logical functions entered into their respective columns. Of course, we must tell the logical operation where the desired inputs are found by using specific cell numbers. This is done by putting the cell reference for the appropriate input at the proper position (or multiple positions) in the equation itself. It would be much easier and less susceptible to errors if we can simply tell the cell containing the operation where the proper inputs are without reediting the equation. This is something that can be done by expressing the logical operations as Excel Functions. If we can do that, then we can simply call the function by name (not by entering the equation) and assign the appropriate signals as passed input parameters to the function. The passed parameters just need to be entered once in the correct order in the function call.

Let us demonstrate this by example using the simple selector circuit in Figure 3-1.

This example only used two kinds of gates; an INVERTER and a NAND gate. Recall, the NAND was used for both the AND and the Invert-OR function. We will use the Visual Basic Editor to create "Functions" for each of these gates.

We start with the inverter. This function has one passed parameter, specifying the input to the gate. First, we need to set the Excel workbook to have Macros enabled. To do this, we open the workbook and Save As "Excel Macro-Enabled Workbook". In the directory, it creates an icon of the workbook with an exclamation mark.

To create the Function, the Visual Basic Editor must be enabled. This is done from the open spreadsheet by selecting [ALT]F11. Once the Editor opens, select "Module" under the "Insert" tab on the top of the page. A blank editing page will open. Let us call the function Invertgate. Enter the following in the editor:

```
Function Invertgate(x)
    If x = 1 Then
        Invertgate = 0
    Else
        Invertgate = 1
    End If
End Function
```

This is the Visual Basic equivalent of the invert equation we have been using in the spreadsheets up to now. One parameter (x) gets passed to it and one value is returned. (Consider adding an error reporting feature to this function to ensure that the entered data is only a 1 or 0).

Now we will create the function for the two-input NAND gate. Let us call it NANDGATE2. By referring to the number of inputs in the function name, this method of labeling will allow us, in the future, to uniquely identify additional gates with more inputs. The function is defined as:

```
Function NANDGATE2(x, y)
    If x = 1 And y = 1 Then
        NANDGATE2 = 0
    Else
        NANDGATE2 = 1
    End If
End Function
```

At this point, the Function can be saved and we can return to our spreadsheet by typing [ALT]Q.

Chapter 3: *Simulation of Combinatorial Logic Circuits* | 43

A Selector spreadsheet is then created which uses Function Calls. It looks like the following:

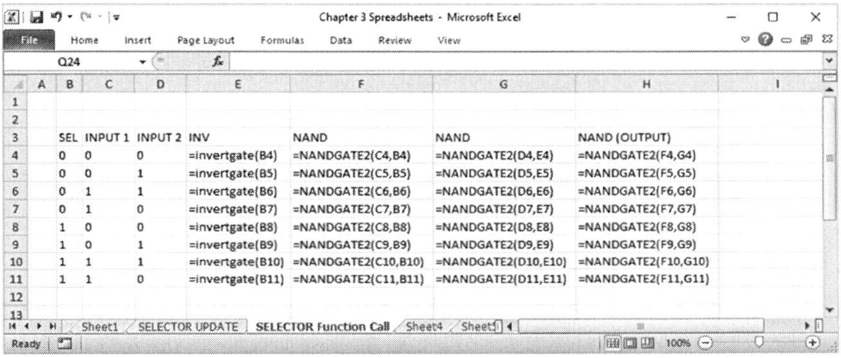

	A	B	C	D	E	F	G	H
3		SEL	INPUT 1	INPUT 2	INV	NAND	NAND	NAND (OUTPUT)
4		0	0	0	=invertgate(B4)	=NANDGATE2(C4,B4)	=NANDGATE2(D4,E4)	=NANDGATE2(F4,G4)
5		0	0	1	=invertgate(B5)	=NANDGATE2(C5,B5)	=NANDGATE2(D5,E5)	=NANDGATE2(F5,G5)
6		0	1	1	=invertgate(B6)	=NANDGATE2(C6,B6)	=NANDGATE2(D6,E6)	=NANDGATE2(F6,G6)
7		0	1	0	=invertgate(B7)	=NANDGATE2(C7,B7)	=NANDGATE2(D7,E7)	=NANDGATE2(F7,G7)
8		1	0	0	=invertgate(B8)	=NANDGATE2(C8,B8)	=NANDGATE2(D8,E8)	=NANDGATE2(F8,G8)
9		1	0	1	=invertgate(B9)	=NANDGATE2(C9,B9)	=NANDGATE2(D9,E9)	=NANDGATE2(F9,G9)
10		1	1	1	=invertgate(B10)	=NANDGATE2(C10,B10)	=NANDGATE2(D10,E10)	=NANDGATE2(F10,G10)
11		1	1	0	=invertgate(B11)	=NANDGATE2(C11,B11)	=NANDGATE2(D11,E11)	=NANDGATE2(F11,G11)

Figure 3-6: Selector Circuit Spreadsheet using Function Calls

In Columns E through H, we only need to enter Row 4 and then copy down the remaining ones. Figure 3-7 shows that the data view of the spreadsheet is the same as we had before (compare to Figure 3-3).

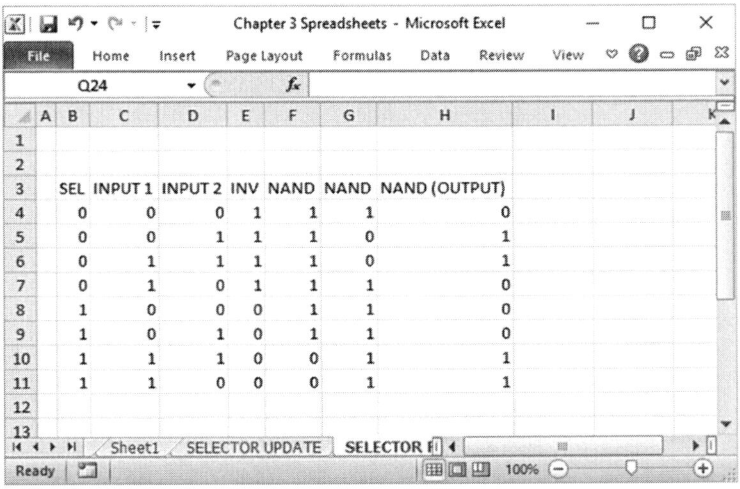

	A	B	C	D	E	F	G	H
3		SEL	INPUT 1	INPUT 2	INV	NAND	NAND	NAND (OUTPUT)
4		0	0	0	1	1	1	0
5		0	0	1	1	1	0	1
6		0	1	1	1	1	0	1
7		0	1	0	1	1	1	0
8		1	0	0	0	1	1	0
9		1	0	1	0	1	1	0
10		1	1	1	0	0	1	1
11		1	1	0	0	0	1	1

Figure 3-7: Data View of Selector Circuit Spreadsheet using Function Calls

44 | Digital Circuit Simulation Using Excel

Let us do one further example. In this case, we will take an arbitrary truth table, design the combinatorial circuitry, perform the simulation, and verify that the implementation is correct.

This circuit consists of four inputs and two outputs as described in Table 3-1.

INPUT A	INPUT B	INPUT C	INPUT D	OUTPUT X	OUTPUT Y
0	0	0	0	1	0
0	0	0	1	1	0
0	0	1	0	0	0
0	0	1	1	0	1
0	1	0	0	0	0
0	1	0	1	0	0
0	1	1	0	0	0
0	1	1	1	0	0
1	0	0	0	1	0
1	0	0	1	0	1
1	0	1	0	0	0
1	0	1	1	1	1
1	1	0	0	0	0
1	1	0	1	0	0
1	1	1	0	0	0
1	1	1	1	0	1

Table 3-1: Four-Input, Two-Output Example Truth Table

Implementing this as a sum of products, the following equations are obtained:

$$\sim A \cdot \sim B \cdot \sim C \cdot \sim D + \sim A \cdot \sim B \cdot \sim C \cdot D + A \cdot \sim B \cdot \sim C \cdot \sim D + A \cdot \sim B \cdot C \cdot D = X$$

$$\sim A \cdot \sim B \cdot C \cdot D + A \cdot \sim B \cdot \sim C \cdot D + A \cdot \sim B \cdot C \cdot D + A \cdot B \cdot C \cdot D = Y$$

There are methods that exist to see if the logic can be reduced (Karnaugh Maps, Quine-McCluskey Method, etc.), however those techniques are not covered here. Actually, by observation, we can see that the first two terms for X can be combined and reduced to:

$$\sim A \cdot \sim B \cdot \sim C \cdot \sim D + \sim A \cdot \sim B \cdot \sim C \cdot D = \sim A \cdot \sim B \cdot \sim C$$

But for the moment, let us simply implement this truth table from the original sum of products equations derived from Table 3-1. Each product term corresponds to a row in the table where the respective output is a one.

The circuit implementation is shown in Figure 3-8.

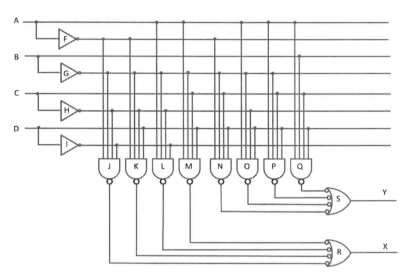

Figure 3-8: Four-Input, Two-Output Example Schematic

Clearly, for this example, another gate must be added to the family of functions. This would be the four-input NAND gate.

The equation is very similar to the two-input gate. Its Visual Basic function definition is:

```
Function NANDGATE4(w, x, y, z)
    If w = 1 And x = 1 And y = 1 And z = 1 Then
    NANDGATE4 = 0
    Else
        NANDGATE4 = 1
    End If
End Function
```

46 | Digital Circuit Simulation Using Excel

Figure 3-8 shows the inputs as labeled A, B, C, and D. The gates are labeled F through S. The outputs X and Y are the outputs of Gates R and S.

The spreadsheet is very wide, the formula view is shown in Figure 3-9 (showing Columns A to J), Figure 3-10 (showing Columns K to O), and Figure 3-11 (showing Columns P to S).

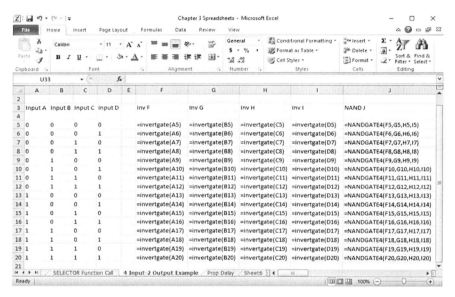

Figure 3-9: Four-Input, Two-Output Example Truth Table, Columns A to J

Chapter 3: *Simulation of Combinatorial Logic Circuits* | 47

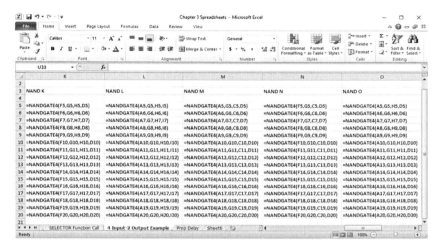

Figure 3-10: Four-Input, Two-Output Example Truth Table, Columns K to O

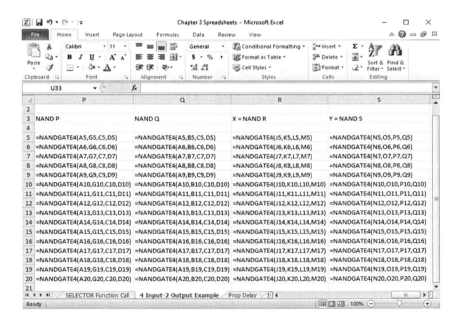

Figure 3-11: Four-Input, Two-Output Example Truth Table, Columns P to S

48 | Digital Circuit Simulation Using Excel

Columns A to D of the spreadsheet show the sixteen input combinations. The pattern progresses in a binary count. These columns were entered manually for each row, but a pattern could have been generated that could be copied down. **(How could this have been done using methods previously introduced?)**

Because of this spreadsheet's symmetry, for the four inverters, only Cell F5 needs to be explicitly entered. It is first copied to the right to Column I then the four columns are copied down for the remaining 15 rows.

Cells in Row 5 for Columns J through S each had to be entered manually because each had a unique input combination. Of course they can be copied down for the remaining rows below Row 5.

The data view of the spreadsheet is shown in Figure 3-12.

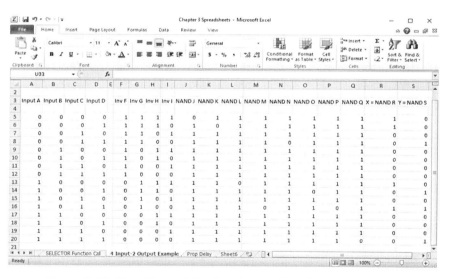

Figure 3-12: Data View of the Four Input – Two Output Example Truth Table.

By comparing Columns R (output X) and S (output Y) of Figure 3-12 with the two rightmost columns of Table 3-1, we see that our simulation verifies that the design implementation is functionally correct.

Additional information on how to use Visual Basic functions in spreadsheets is given in Reference [7].

Additional examples of circuits implemented with basic gates can be found in References [1], [3], [4], [5], [6], and [9]. Some of these examples include such functions as adders (full and half), multiplexers, demultiplexers, and decoders.

NAND and NOR implementations for combinatorial logic designs are covered in References [5], and [8].

There are many sources for information on logic reduction techniques like Karnaugh Maps or the Quine McCluskey Method such as References [3], [4], [5], [6], [8], and [9].

Implementation of logical functions as Sum-Of-Products or Product-Of-Sums can be found in Reference [3]. Reference [3] also has an introduction to timing diagrams.

CHAPTER

4

Memory Devices

In this chapter, the structure and simulation of memory elements are discussed. These are devices that can be set to a 1 or a 0 by some action defined for its inputs. When the inputs return to a quiescent state, the device will hold its value so that it can be retrieved later on.

The devices covered here are the latch and the flip flop.

R-S Latch

The simplest memory element is the R-S latch, which is composed of two NAND gates configured as shown in Figure 4-1.

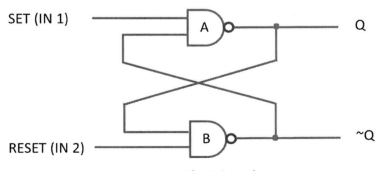

Figure 4-1: The R-S Latch

Both the Set and Reset inputs are normally kept in a "1" or high level to retain the state of the latch. Let us assume for a moment that Q=0 and ~Q=1. If Set and Reset remain as 1, then Gate A sees 1's on both of its inputs which maintains its output of 0. Gate B has inputs of 0 and 1, giving an output of 1. So the state of the latch is maintained as long as both inputs are 1. Naturally, the same result is observed if the original state of the latch were Q=1 and ~Q=0. That is, the state of the latch persists as long as both inputs remain at 1.

Starting again in the Q=0 and ~Q=1 state, if a low pulse is presented at Reset, Q and ~Q still remain in the (0,1) state since Gate B will still output a 1. But let us see what happens if a low pulse is presented on Set. If Set goes to 0 at the input of Gate A, Q transitions to a 1. This gives Gate B a 1 on both of its inputs, making ~Q=0. The ~Q signal is fed back to Gate A, maintaining the 1 at Q. So the latch has changed its state to (1,0). When both Set and Reset return to 1, the new Q=1 and ~Q=0 state is retained.

The only requirement that needs to be imposed on the Set/Reset pulses to ensure that the state of the latch is validly changed is that they persist longer than the propagation delay through Gates A and B. **(Can you see why? We will look into this issue shortly)**.

Having both SET and RESET go to a value of zero at the same time is problematic. If this should occur, both outputs go to a 1 state. But if both SET and RESET return to a 1 at the same time, the state of the latch cannot be predicted. So this situation is generally avoided. However, if only one of the inputs returns to a 1 long enough for the gates to settle to a (1,0) or (0,1) state, then we have a predictable situation. If SET returns to 1 first, Q becomes 0 and ~Q becomes 1. Whereas if RESET returns to 1 first, Q becomes 1 and ~Q becomes 0. From this point, operation proceeds normally.

Let us look into the behavior of the latch in a little more detail by simulating it in Excel. A straightforward implementation of Figure 4-1 is given in Figure 4-2.

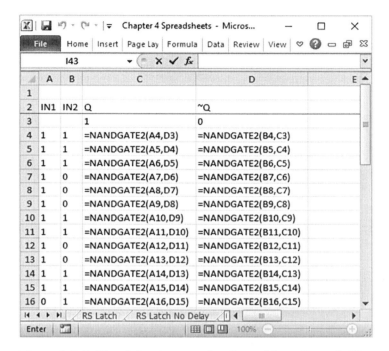

Figure 4-2: R-S Latch Implementation Spreadsheet Formula View

IN1 (SET), IN2 (RESET), Q, and ~Q are in Columns A, B, C, and D respectively.

The input for Columns A and B are simply a sequence chosen to demonstrate the latch's operation. It does not include the situation where both inputs are low at the same time. This will be looked at later on.

Columns C and D define the output of each of the latch's constituent gates. Row 3 established an arbitrary initial condition for the latch where Q=1 and ~Q=0. This is necessary since the current state in the simulation depends on the previous state. Notice, for example, that cells in Row 4 reference cells in Row 3.

54 | Digital Circuit Simulation Using Excel

Starting with Row 4, we have simply employed the NAND operation between the current state of the input (IN1 for the Q Gate and IN2 for the ~Q Gate) and the output of the complementary NAND gate. Row 3 of the ~Q Gate is selected for the input to the Q Gate in Row 4, since Row 3 represents the present state of ~Q with respect to Row 4. Row 4 will generate the next state.

The data view for the spreadsheet is given in Figure 4-3.

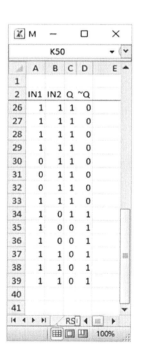

Rows 1-25   Rows 26-39

Figure 4-3: Data View of the Spreadsheet for the R-S Latch Operation

Let us review what has happened. Row 3 established an initial state for the latch with Q=1 and ~Q=0. Starting in Row 4, we have the inputs to the latch in the (1,1) condition so the state of the latch does not change. The (1,1) input persists through Row 6 and the state of the output is maintained.

In Row 7, a 0 is presented on IN2 in an attempt to change the latch's output state to (0,1). But we notice something a little unusual. Both of the outputs go to the 1 state in Row 7. A little observation shows why this is the case. When IN2 goes to a 0, ~Q will go to a 1 in response. But Q will only go to a 0 after the 1 at ~Q has been established on the output of its NAND gate (the gate represented by Column D) and can then be interpreted as the "previous" ~Q to the NAND gate controlling Q in Column C. So it is indeed valid to see both outputs 1 for a brief period of time corresponding to the propagation delay of the NAND gates.

The propagation delays for simple standard gates (74LS00 series for example) are generally less than 20 ns. For the sake of predictable operation, specifically with respect to timing considerations, in a circuit design, the signals controlling these circuits (IN1 and IN2 in this case) should not be switching faster than or near that rate.

Continuing after Row 7, Row 8 indicates that IN2 is still a 0, but now ~Q has been fed back to the input of Q's NAND gate, making Q go to 0. At Row 10, IN2 goes high and the outputs remain at their current values. At Rows 12 and 13, we again pulse IN2 low and, as expected, the output states do not change.

At Row 16, we change the state of the latch by pulsing IN1 to 0. As before in Row 7, both outputs are 1. When the Q output gets fed back to the input, in Row 17, ~Q changes to a 0.

At Row 19, both inputs are 1 and the outputs retain their states.

Again we pulse IN1 low at Rows 22 and 23, then again in Rows 30 to 32. We do not expect to see a state change and we do not.

Finally in Row 34, we pulse IN2 low to change the state. Row 34 shows the temporary (1,1) state, and after that the latch responds as expected.

In Figure 4-4, we show the signals in timing diagrams following the input sequence of Figure 4-3.

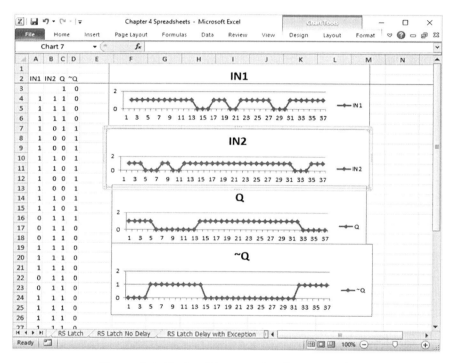

Figure 4-4: Timing Diagrams for the R-S Latch

Let us digress a moment and consider how propagation delays factor into the simulation spreadsheets. This is an important point to consider when we have circuits that have signals involved in feedback as in this situation with the R-S latch. If we were to go back to Figure 4-2 and change the expressions for Q and ~Q in Row 4, for example, so that we are only using values from the current row, rather than the previous row, the change would be:

Cell C4: NANDGATE2(A4,D4) rather than NANDGATE2(A4,D3)

Cell D4: NANDGATE2(B4,C4) rather than NANDGATE2(B4,C3)

Excel would flag a "Circular Reference Warning." This is because Excel would see that the changed state of the circuit gets fed back in the same row to change the inputs that are generating it. In the physical realization of the circuit, because of circuit propagation delays, this is not an issue. We just need a way to work this into our spreadsheets. This is done by simply letting the propagation delay in a feedback circuit be represented by using an earlier row in the spreadsheet. Therefore we took the gate's output from the previous row of the spreadsheet as was shown above.

For simple random logic circuits where there is no memory or feedback and where we are not trying to simulate transient states arising from a switching hazard or race condition, we can generally avoid accounting for propagation delays. We will discuss simulating transient switching states beginning in Chapter 7.

This highlights the difference between the random logic circuits discussed in the previous chapters and those that have memory. In the previous analyses, the spreadsheets represented a static verification of the truth tables of the design. The output was only dependent on the present inputs (after transients states have settled out). Here, we see the spreadsheet simulating a dynamic situation (i.e. each subsequent spreadsheet row represents the next sequence in time) where outputs depend on both the current inputs as well as the previous state of the circuit. Again, this is a consequence of having memory elements and will be covered more thoroughly beginning in Chapter 5.

In order to highlight one other characteristic of the R-S latch, let us look again at the situation where we had both outputs momentarily equal to 1 at the same time. We mentioned that the occurrence of the transient (1,1) state puts a restriction on the R and S inputs. Specifically, in a transition that generates this state (e.g. (0,1) to (1,1)

58 | Digital Circuit Simulation Using Excel

to (1,0)), each row in this area represents a duration of time equal to the propagation delay through a gate. Going from (0,1) to (1,1) represents the propagation through one gate, while the (1,1) to (1,0) transition indicates the propagation through the second. So, we can see that in order to reliably control the state of the latch, the Set or Reset input must persist for two rows in the spreadsheet (i.e. the propagation delay for two gates). If either signal were only present in one row, the output would oscillate as shown in Figure 4-5.

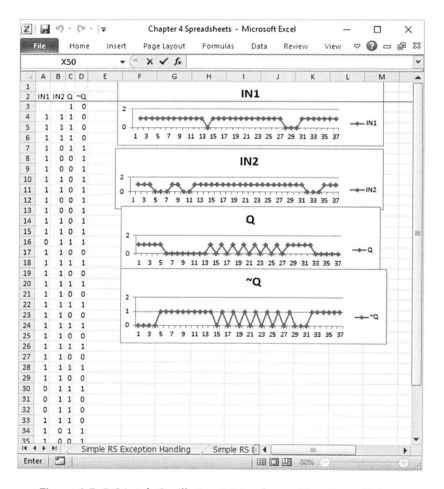

Figure 4-5: R-S Latch Oscillation Arising from a Short Input Pulse

Figure 4-5 is the same as Figure 4-4, except that the IN1 pulse introduced in Row 16 is only present for that one row. The outputs are shown oscillating until a valid input signal is presented in Rows 30 to 32. **(Can you see why the short input pulse creates oscillations?)** Therefore, we must ensure that the triggering signals for the latch last for at least two rows; assuming, of course, that one row's duration equals the propagation delay through one gate.

The previous spreadsheet analysis (Figure 4-4) avoided the situation where IN1 and IN2 are both 0 at the same time. We know that if this should occur, both outputs will go to 1. The problem occurs when we release the inputs from this state.

As mentioned earlier, if the inputs go directly from a (0,0) state to either a (0,1) or (1,0) state, there really is no issue. The latch will be driven to the Q=1/~Q=0 state for the former and Q=0/~Q=1 for the latter. The problem occurs if both are released to 1 at approximately the same time. In that situation, assuming the propagation delay of both gates is the same or very close, there will be 1's at both of the NAND gates. This will make both gates output 0. But the 0's will be fed back and drive the outputs to 1's again. It would appear that an oscillation would be present on the outputs. Figure 4-6 shows how the spreadsheet responds.

60 | Digital Circuit Simulation Using Excel

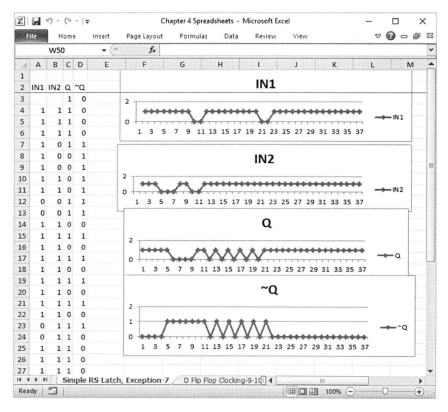

Figure 4-6: R-S Latch with (0,0) Inputs

Up until Row 12, the latch is undergoing standard operation. Both inputs start in the (1,1) state. Rows 7 through 9 show IN2 pulsing low to change the output state. Then at Rows 12 and 13, both inputs go to 0. Notice that the outputs during this time are both 1. However, when the inputs go back to 1 starting at Row 14, Q and ~Q begin to oscillate. The oscillations continue until the 0 on IN1 (Row 23) brings it back to a stable state. This is similar to the situation we encountered when the R or S pulse persisted for less than the propagation delay of two gates.

In practice, the oscillations in the latch may stop on their own after a while. Or if the propagation delays of the gates are different enough, we may see limited or no oscillation. However, it is not possible to predict the eventual state of the latch until we drive it to a known state.

The designer must take care to account for this condition. Oscillations in the spreadsheet simulation will be able to highlight this situation to the designer.

Because of its simplicity, we will not create a new function for the R-S latch. To do this, two functions would be created: one for the Q output and one for the ~Q output. But each of these would simply use the already existing NANDGATE2 function.

D Flip Flop

Flip flops are similar to latches in that they are memory elements that store binary values. The main difference is that the outputs of latches asynchronously change when the inputs to them (SET and RESET) change and drive a transition to a new state. While D flip flops (and most other flip flops) also support an asynchronous set and reset (or clear), they also have a data input that only affects the state of the output in association with a clock.

There are many advantages to this. One of the most important is that a circuit comprised of several flip flops that are synchronously clocked has all outputs change at the same time and will not have updating data rippling through the circuit dependent on the propagation delay of its components. This makes accommodating race conditions and switching hazards much more manageable and predictable. We will return to this topic when we discuss asynchronous and synchronous circuits.

The circuit representation for the D flip flop is shown in Figure 4-7.

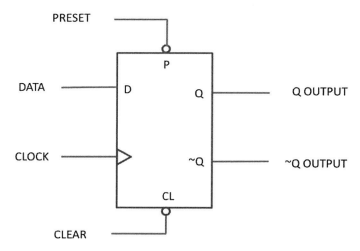

Figure 4-7: Circuit Representation of the D Flip Flop

On the rising edge of the CLOCK input, the value present on DATA becomes the state of the Q OUTPUT. The ~Q OUTPUT takes on the opposite value of Q. DATA must meet the specified setup time (minimum time for the signal on D to be stable prior to the clock's rising edge) and hold time (minimum time for the signal on D to be stable after the clock's rising edge). For the LS family of devices, the setup time is around 20 ns while the hold time is around 5 ns.

PRESET, an active low signal (indicated by the bubble on the diagram), is used to set the Q OUTPUT high. It is asynchronous and so operates independently of the clock. CLEAR, also active low, forces the Q OUTPUT low. It is also independent of the clock. Having both PRESET and CLEAR low simultaneously will generally cause both Q and ~Q to go high. However, the device manufacturers do not recommend this situation and won't guarantee stability once the simultaneously active PRESET and CLEAR signals are removed. We will avoid that situation in our spreadsheet simulation by adopting an arbitrary convention that when the PRESET and CLEAR inputs are both low, both the Q and ~Q outputs will be set to -1. Since a -1 is an invalid logic state, it will be an obvious indication that an invalid sequence was present on the inputs and must be corrected. Also, if we generate timing diagrams from the spreadsheet, the presence of a -1 appearing in a normal 0,1 output signal will be quite apparent. In this simulation, the -1 will persist until the flip flop is driven into a valid state.

The flip flop symbol is standard and is normally shown without any labeling on the inputs and outputs. The exception is that the triangle on the clock is usually shown.

Let us now address how the D flip flop is implemented in a spreadsheet. A first attempt at simulating this device gives the following formula for the Q output:

=IF(AND(PRESET=0,CLEAR=0),-1,IF(PRESET=0,1,IF(CLEAR=0,0,IF(AND(CLOCK=1,CLOCKX=0),DATA,QX))))

The six inputs are expressed by name for this description. Naturally, in the spreadsheet, the names are replaced by the identifiers of the cells that provide the appropriate parameter values.

The six inputs are:

    PRESET – The current value for the Preset input

    CLEAR – The current value for the Clear input

    CLOCK – The value of the CLOCK signal at the place where we look for a transition to 1 to clock D to Q

    CLOCKX – The value of the CLOCK signal at the previous spreadsheet row to the one used for CLOCK

    DATA – The signal at the DATA input

    QX - The value of the Q output at the previous row of the spreadsheet.

The first condition checked is for the error situation that would occur if both PRESET and CLEAR are 0 at the same time. This condition has the highest priority. If it occurs, our formula has the Q output set to a -1 rather than 0 or 1. The -1 notification will continue to be output until the flip flop is put into a valid state by

1. PRESET going low with CLEAR high;
2. CLEAR going low with PRESET high;
3. Setting the output to the state on DATA at a positive transition of CLOCK.

Given that the PRESET=0 and CLEAR=0 error condition does not exist, the case where either PRESET or CLEAR is 0 is checked since either situation would override the clocked input. If PRESET is 0, then Q would be set to 1 and if CLEAR is 0 then Q would immediately transition to 0.

If PRESET and CLEAR are both 1, then the clock is checked to see if a positive transition has occurred. This is done by examining sequential states of the clock (i.e. consecutive spreadsheet rows) to see if a 0 state was followed by if 1 state. If that occurred, then a positive edge occurred and Q takes on the value appearing on DATA at the clock transition.

At this point, if there was no positive transition on the clock, then Q simply retains its previous value even when the previous value is the -1 error indication.

A formula for ~Q is also generated. The argument can be made that ~Q is simply the inversion of Q (except, of course, for the -1 error condition). However, we will go ahead and generate a separate formula for ~Q that does not rely on Q being evaluated. The reason for this is that in a particular design, the Q output might not be required when the ~Q output is used. In that case, there is no need to create a spreadsheet column for Q simply to generate ~Q as its inverse.

Based upon the current formula for Q, stand-alone formula for ~Q is:

=IF(AND(PRESET=0,CLEAR=0),-1,IF(PRESET=0,0,IF(CLEAR=0,1,
    IF(AND(CLOCK=1,CLOCKX=0),IF(DATA=0,1,0),~QX),))))

The check for the PRESET=0, CLEAR=0 error is the same as for Q. If it occurs, the -1 error indication will be generated.

Next either PRESET or CLEAR (but not both) being 0 is checked. But in this case, the active low PRESET will make the ~Q output 0 while the CLEAR will make it 1, opposite of Q.

The occurrence of a positive edge transition on the CLOCK is checked. If it occurred, then we must output a 0 if DATA was 1 or a 1 if DATA was 0.

If none of these situations occurred, ~Q is simply equal to its previous value.

Figure 4-8 and Figure 4-9 show a spreadsheet formula view which is used to test the D flip flop.

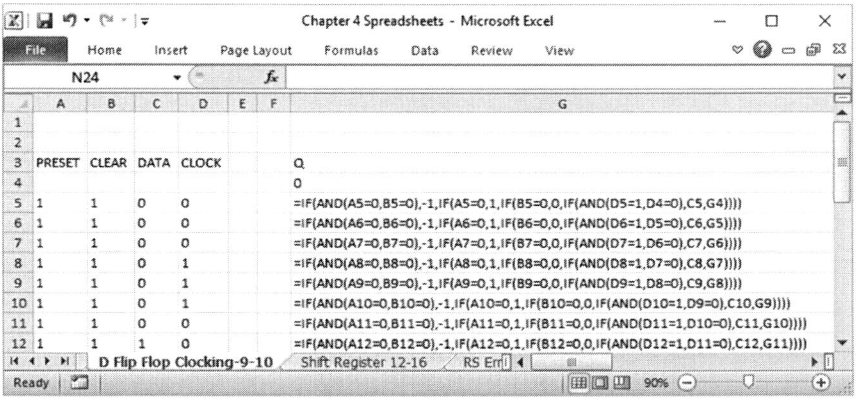

Figure 4-8: Formula View of the Spreadsheet for Testing the D Flip Flop, Columns A through G

66 | Digital Circuit Simulation Using Excel

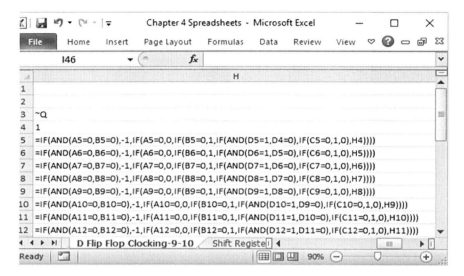

Figure 4-9: Formula View of the Spreadsheet for Testing the D Flip Flop, Column H

The formulas for deriving Q and ~Q are in Columns G and H while the inputs are given in Columns A through D. Q and ~Q are initialized in Row 4.

A simple scenario for testing how the outputs respond to clocking data is given in Figure 4-10 with comments highlighting the major points of behavior. At the moment, we are assuming that the Data to Q clocking is done at the time the CLOCK input transitions to a 1. Also note that we are not yet testing the response of the flip flop with respect to active PRESET, active CLEAR, or the error condition when they are both active. We will return to that shortly since we will see an additional modification is warranted in that area.

Figure 4-10: Clocking Behavior of D Flip Flop

The formulas shown here appear to work fine. However, let us look into the issue of when to update Q and ~Q in these formulas based on a positive clock transition. If we are not careful, we may see some unexpected results. A problem can be demonstrated very clearly if we consider a simple shift register design using the D flip flop. Recall that a shift register moves data down successive elements of the register at each occurrence of a clocking signal. Keeping with the D

68 | Digital Circuit Simulation Using Excel

flip flop implementation, let us say that the rising edge of the clock triggers the shift to the next element.

In practice, a shift register is easily implemented using a series of D flip flops as shown in Figure 4-11. We will not show PRESET and CLEAR since they will not be used. Assume they are connected to a logical 1 signal. In the spreadsheet the PRESET and CLEAR values are set by referring them to a cell with a value of 1 or by simply setting their values to 1 in the formulas. In a physical circuit they are usually connected to a pull-up resistor.

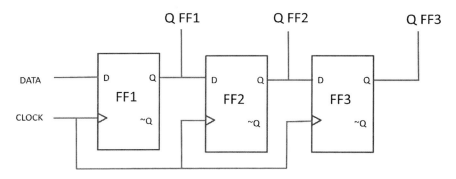

Figure 4-11: Shift Register Implementation Using D Flip Flops

The expected behavior would be to see a signal (actually it would be whatever was on DATA at the rising edge of CLOCK) clocked into FF1 and appear at output Q of FF1. On the next rising edge of the clock, the original signal would appear at Q of FF2 while the current value of DATA is clocked into FF1. On the next clock edge, the original signal would appear on Q of FF3.

Using our spreadsheet, we might consider that the correct formula view is as shown in Figure 4-12 and Figure 4-13 (since the equations are rather long, this view was split into two figures). However, this leads to the data view in Figure 4-14. DATA starts off as 0 then in Row 7 it becomes a 1. The rising edge of CLOCK (Rows 8 and 9) transfers DATA to Q of FF1 as expected. However, the data is also transferred to Q of FF2 and Q of FF3 on the same clock transition (Rows 9 through 16 of Columns D and E).

It is clear what happened if we think about what the spreadsheet does versus what actually happens in the physical circuit. When we calculate the spreadsheet row by row (consider Rows 8 and 9), each column is calculated without taking signal propagation delays into account. So when the positive clock transition happened on Rows 8 to 9, the output Q of FF1 was changed immediately at that time and the updated output was available to the columns that followed. Those following columns represent subsequent circuitry. Therefore, when FF2 saw the same clock edge, the Q output of FF1 was already high and it clocked through the high level as well. The same thing happened to FF3 since FF2's output changed immediately.

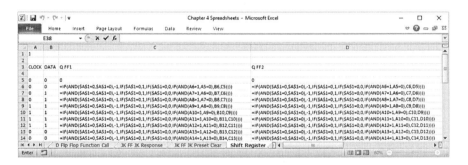

Figure 4-12: Spreadsheet Formula View of the Initial Attempt to Implement a Shift Register (Columns A to D)

In reality, what happens is that there is delay through FF1. So although FF2 sees the same edge, the updated value of Q of FF1 has not been propagated through the flip flop and has not met the setup time for FF2.

70 | Digital Circuit Simulation Using Excel

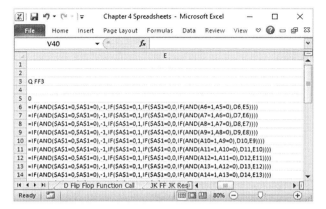

Figure 4-13: Spreadsheet Formula View of the Initial Attempt to Implement a Shift Register (Column E)

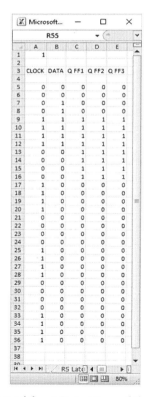

Figure 4-14: Spreadsheet Data View of the Initial Attempt to Implement a Shift Register

To work around this issue, there is nothing we need to change in the spreadsheet function definitions. All we need to do is account for the propagation delays through the flip flops. We must allow the data to be transferred to Q in response to the rising clock edge, but it should not be available at Q at that same clock edge to the circuitry that is downstream. All we need to do is to ensure that Q does not change until the next spreadsheet row after the rising edge transition on its clock.

Figure 4-15 shows the updated spreadsheet for the FF1 and FF2 flip flops. There is an analogous change for FF3 which is not shown in the figure.

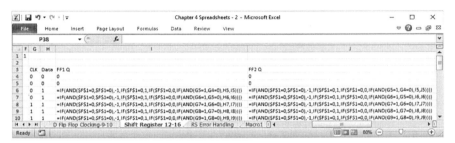

Figure 4-15: Corrected Spreadsheet Formulas for the Shift Register Implementation (Columns F through J)

All that was done was move the search for the clock transition up one row. Consider Row 10 in Column I. It now says to consider the clock transitions between Rows 9 and 8 (i.e. it specifies G9 and G8). In the first spreadsheet it looked for the transition occurring between Rows 10 (the current row) and 9. The change just keeps Q and ~Q from changing immediately and being available to subsequent circuitry. This is done for all three flip flops. It delays one row in the spreadsheet. The new data view and timing diagrams are shown in Figure 4-16 indicating that the expected behavior was achieved.

72 | Digital Circuit Simulation Using Excel

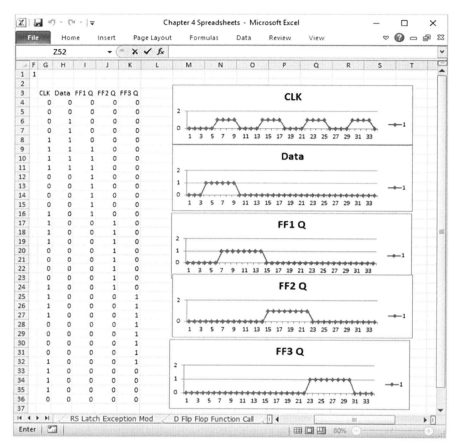

Figure 4-16: Data View of the Corrected Spreadsheet for the Shift Register Implementation

Now the simulation will be examined to see how the flip flop responds to PRESET and CLEAR. Up to now, the spreadsheet simulations for these inputs have just been held to a value of 1.

Since we have just discussed the issue of data to output delay, let us think about the same timing consideration with respect to PRESET and CLEAR.

Chapter 4: *Memory Devices* | 73

There are two issues to consider. The first deals with propagation delay through the flip flop. The way this type of flip flop is generally designed, PRESET and CLEAR affect the state of Q and ~Q independently of CLOCK. The internal design of the flip flop indicates that PRESET and CLEAR have the same propagation delay as the clock with regard to affecting Q and ~Q. Typical manufacturer's data sheets bear this out as well. So in the equations for Q and ~Q, the values for PRESET and CLEAR should be taken from the previous row.

Given this decision, we then need to consider how a PRESET or CLEAR occurring in the vicinity of a rising clock edge affects the flip flop's output. In our simulation, a positive edge on the clock is seen as 0 occurring in one row of the spreadsheet followed by a 1 in the next row. Remember, the decision to update Q based on the clock is delayed until the next row after the clock signal rises to a 1 representing the propagation delay through the flip flop.

Because PRESET and CLEAR override the clocking function, we require that a PRESET or CLEAR occurring in the interval when the clock goes to a 1 must be reflected in the output regardless of what value on the DATA input was clocked in. That is Q and ~Q will follow the PRESET or CLEAR functionality rather than the clocked value from D at this time. This is unambiguous since PRESET or CLEAR will have occurred after the clock transition. So the PRESET/CLEAR override is consistent with taking its value from the previous row in the spreadsheet.

It can be argued whether PRESET and CLEAR should also override the clocked data if the pulse occurs in the interval just before the clock transition where it is still zero. This would depend on how the flip flop is internally designed. Many schematics showing the structure of the D flip flop indicate that an active PRESET and CLEAR that becomes inactive in the interval before the clock transition will not affect the clocking from DATA to Q and ~Q. We will adopt that convention here. Note that if this behavior needs to be changed

74 | Digital Circuit Simulation Using Excel

based on a different flip flop design, it is easy to modify our D flip flop equations to allow the clocking to be overridden in the interval before the clock transition.

To be clear, Figure 4-17 shows how the design will be implemented.

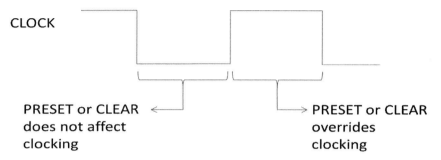

Figure 4-17: PRESET and CLEAR Operations with Respect to the Clock

Considering all of these situations, we arrive at the following spreadsheet formula views for the D flip flop (Figure 4-18 and Figure 4-19):

	A	B	C	D	E F	G
1	PRESET	CLEAR	DATA	CLOCK		Q
2	1	1	0	0		0
3	1	1	0	1		0
4	1	1	0	1		=IF(AND(A3=0,B3=0),-1,IF(A3=0,1,IF(B3=0,0,IF(AND(D3=1,D2=0),C3,G3))))
5	1	1	0	1		=IF(AND(A4=0,B4=0),-1,IF(A4=0,1,IF(B4=0,0,IF(AND(D4=1,D3=0),C4,G4))))
6	1	1	0	0		=IF(AND(A5=0,B5=0),-1,IF(A5=0,1,IF(B5=0,0,IF(AND(D5=1,D4=0),C5,G5))))
7	1	1	0	0		=IF(AND(A6=0,B6=0),-1,IF(A6=0,1,IF(B6=0,0,IF(AND(D6=1,D5=0),C6,G6))))
8	0	1	0	0		=IF(AND(A7=0,B7=0),-1,IF(A7=0,1,IF(B7=0,0,IF(AND(D7=1,D6=0),C7,G7))))
9	0	1	0	1		=IF(AND(A8=0,B8=0),-1,IF(A8=0,1,IF(B8=0,0,IF(AND(D8=1,D7=0),C8,G8))))
10	1	1	0	1		=IF(AND(A9=0,B9=0),-1,IF(A9=0,1,IF(B9=0,0,IF(AND(D9=1,D8=0),C9,G9))))

Figure 4-18: Spreadsheet Formula View for PRESET and CLEAR Updates Columns A Through G

Chapter 4: *Memory Devices* | 75

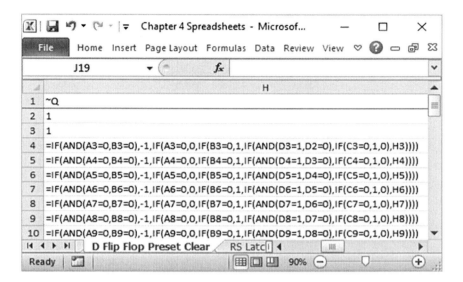

Figure 4-19: Spreadsheet Formula View for PRESET and CLEAR Updates Column H

The equations for Q and ~Q are similar.

First the PRESET and CLEAR error condition is checked for their values in the previous row. It there is no error, then the PRESET and CLEAR conditions are separately checked to see if either one is active. If so then Q and ~Q are immediately updated as appropriate.

If no PRESET/CLEAR activity occurred, the clock is checked to see if there was a positive transition. Looking at Row 4, the clock in Rows 2 and 3 are checked (Cells D2 and D3). If D2 is 0 and D3 is 1, then a positive transition occurred. Since we reached this point because neither PRESET nor CLEAR was active, we allow DATA (C3) to be latched at the output. If we did not have a clock transition, we output to Q and ~Q their respective values from the previous row. Figure 4-20 and Figure 4-21 show some scenarios demonstrating the flip flop's behavior in the presence of PRESET and CLEAR signals.

Figure 4-20 shows the basic functionality as well as the detection of both signals being active at the same time. Figure 4-21 tests the flip flop's reaction to PRESET and CLEAR in the vicinity of a rising clock edge.

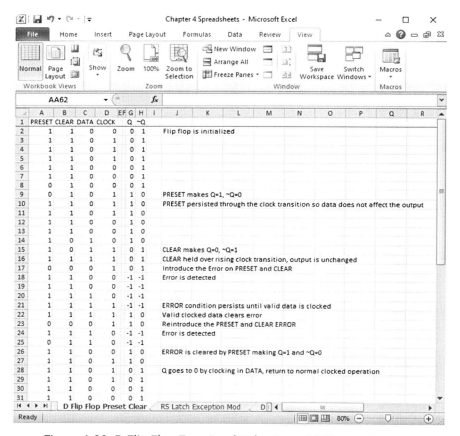

Figure 4-20: D Flip Flop Functional Behavior to PRESET and CLEAR

Chapter 4: *Memory Devices* | 77

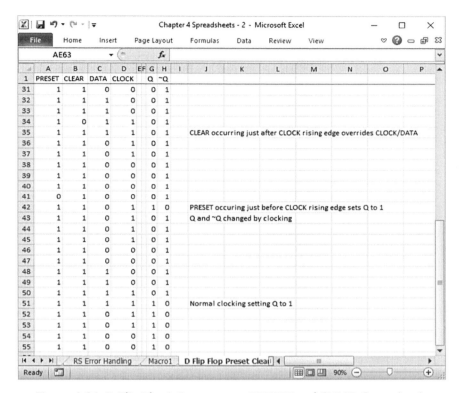

Figure 4-21: D Flip Flop's Response to PRESET and CLEAR Occurring in the Vicinity of CLOCK's Rising Edge

To invoke the D flip flop as a function, we can create a new Visual Basic program that can be added to the library. The function for the Q output uses passed parameters PRESET, CLEAR, DATA, CLOCK CLOCKX, QX where:

PRESET, CLEAR, DATA, CLOCK, and QX are referenced in the previous spreadsheet row

CLOCKX is referenced two rows back.

The program is as follows:

```
Function DFFQ(PRESET, CLEAR, DATA, CLOCK, CLOCKX, QX)
        If PRESET = 0 And CLEAR = 0 Then
        DFFQ = -1
        Else
                If PRESET = 0 Then
                DFFQ = 1
                Else
                        If CLEAR = 0 Then
                        DFFQ = 0
                        Else
                                If CLOCKX = 0 And CLOCK = 1 Then
                                DFFQ = DATA
                                Else
                                        DFFQ = QX
                                End If
                        End If
                End If
        End If
End Function
```

Similarly, a program can be created for the ~Q function. The passed parameters and their respective location in the spreadsheet are the same as those defined for the program DFFQ with three exceptions. First, the parameter NOTQX replaces QX. Second, when there is a valid clock, ~Q takes on the inverted value of DATA. Third, the active PRESET and CLEAR actions for ~Q are in the opposite logical state as Q.

```
Function DFFNOTQ(PRESET, CLEAR, DATA, CLOCK, CLOCKX, NOTQX)
        If PRESET = 0 And CLEAR = 0 Then
        DFFNOTQ = -1
        Else
                If PRESET = 0 Then
                DFFNOTQ = 0
```

```
            Else
                    If CLEAR = 0 Then
                        DFFNOTQ = 1
                    Else
                            If CLOCKX = 0 And CLOCK = 1 Then
                                If DATA = 0 Then
                                    DFFNOTQ = 1
                                Else
                                        DFFNOTQ = 0
                                End If
                            Else
                                    DFFNOTQ = NOTQX
                            End If
                    End If
            End If
    End If
End Function
```

The formula view of the spreadsheet performing some simple tests using the function calls for Q and ~Q is shown in Figure 4-22.

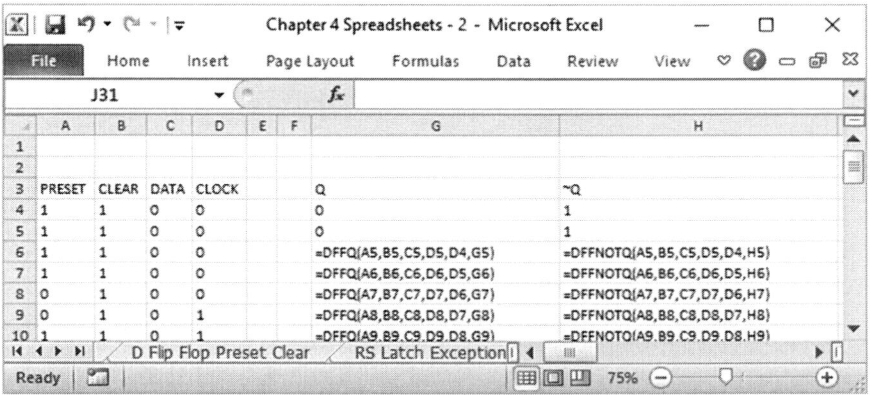

Figure 4-22: Formula View of the Spreadsheet Implementing the D Flip Flop as a Function Call

A data view of the same spreadsheet is shown in Figure 4-23. Some basic tests are executed to show that the expected functionality is achieved. The key events are indicated in Column J. Timing diagrams of the inputs and the Q output are given.

The flip flop operates as expected. Notice that when the error occurred where PRESET and CLEAR were both active, Q and ~Q took on the value of -1. This is clearly observed in the graph showing Q.

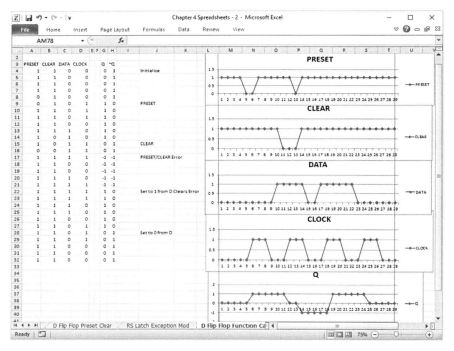

Figure 4-23: Data View of Spreadsheet Implementing the D Flip Flop as a Function Call

J-K Flip Flop

The J-K flip flop has a total of five inputs: PRESET, CLEAR, J, K, and CLOCK. Its outputs are Q and ~Q. The circuit representation is shown in Figure 4-24.

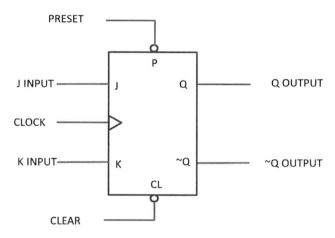

Figure 4-24: Circuit Representation of the J-K Flip Flop

With regard to performance, the J-K flip flop allows all combinations of values for its J and K inputs. The state transitions for the input combinations are shown Table 4-1.

J	K	Next Q	Next ~Q	Comment
0	0	Current Q	Current ~Q	Keeps the Current State
0	1	0	1	Sets Q=0 and ~Q=1
1	0	1	0	Sets Q=1 and ~Q=0
1	1	Current ~Q	Current Q	Toggles the Output

Table 4-1: State Definitions for the J-K Flip Flop

PRESET and CLEAR work the same way as they do for the D flip flop. They operate independently of CLOCK and they both should not be active (low) at the same time. So the error detection incorporated into the D flip flop operation will be retained for the J-K flip flop.

Given that PRESET and CLEAR are both high, the output will follow the J and K inputs as indicated by Table 4-1. Note that in practice, J-K flip flops can be found that are either positive or negative clock edge triggered. Others operate on both edges of the clock where the rising edge stores J and K in a first stage internal latch while the falling edge transfers the information to the Q/~Q output stage. For our simulation, we will just consider positive edge triggered. This is the design implemented in the standard 74109/74LS109 device. This means the behavioral conventions added to the D flip flop are kept for the J-K flip flop. These were:

1. Clocked values from J and K to Q and ~Q will appear at the outputs in the interval (i.e. the spreadsheet row) where the clock has transitioned to a positive value.
2. A PRESET or CLEAR in the interval where the clock has transitioned to a positive value overrides the clocking or J and K.

The Visual Basic Editor functions for the J-K flip flop follow.

The Q output is defined by the function JKFFQ as shown below. The line numbers 1 to 39 are not part of the function; they are included in this listing as a reference for the subsequent explanation.

```
1 Function JKFFQ(PRESET, CLEAR, J, K, CLOCK, CLOCKX, QX)
2       If PRESET = 0 And CLEAR = 0 Then
3           JKFFQ = -1
4       Else
5           If PRESET = 0 Then
6               JKFFQ = 1
7           Else
```

```
 8              If CLEAR = 0 Then
 9                  JKFFQ = 0
10              Else
11                  If CLOCK = 1 And CLOCKX = 0 Then
12                      If J = 0 And K = 0 Then
13                          JKFFQ = QX
14                      Else
15                          If J = 0 And K = 1 Then
16                              JKFFQ = 0
17                          Else
18                              If J = 1 And K = 0 Then
19                                  JKFFQ = 1
20                              Else
21                                  If QX <> -1 Then
22                                      If QX = 0 Then
23                                          JKFFQ = 1
24                                      Else
25                                          JKFFQ = 0
26                                      End If
27                                  Else
28                                      JKFFQ = QX
29                                  End If
30                              End If
31                          End If
32                      End If
33                  Else
34                      JKFFQ = QX
35                  End If
36              End If
37          End If
38      End If
39 End Function
```

Where:

    PRESET is the signal on the PRESET input during the immediately preceding spreadsheet row

CLEAR is the signal on the CLEAR input during the immediately preceding spreadsheet row

J is the signal on the J input during the immediately preceding spreadsheet row

K is the signal on the K input during the immediately preceding spreadsheet row

CLOCK is the signal on the CLOCK input during the immediately preceding spreadsheet row

CLOCKX is the signal on the CLOCK input referenced two spreadsheet rows back

QX is the value for Q during the immediately preceding spreadsheet row

Lines 2 and 3 of the listing check for the error condition where PRESET and CLEAR are both active at the same time. If this occurs, then the -1 error indication is set and the program terminates.

If there is no error condition, then lines 5 and 6 check for an active PRESET. If so, then Q is set to 1 in line 6. If not, then CLEAR is checked (line 8) to see if it is active. If so, Q is reset to 0 in line 9. If neither of these signals is active, then the function proceeds to look for a positive clock transition occurring in the previous two rows (line 11). It does this by seeing if the clock was 0 two rows back and then transitioned to 1 in the previous row. If these conditions are met, then J and K are examined to see if and how Q will be affected by the positive clock transition.

First, line 12 sees if J=0 and K=0. This means there is no change in Q. If this condition is met, then line 13 sets Q to its previous value.

If J=0 and K=1 (line 15) then Q is set to 0 (line 16).

If J=1 and K=0 (line 18) then Q is set to 1 (line 19).

If none of the above cases are met, then J and K both must be 1 which normally invokes a toggle. However, it must also check for Q being -1, which means that there has been an uncleared PRESET=0/ CLEAR=0 error condition. If an error indication has already been set, the -1 state on Q needs to be retained since toggling an error condition does not yield a valid defined state. It should keep the flip flop in an error condition. So line 21 checks the last value of Q. If it was not equal to -1, then it is set to 1 if the previous value was 0 or set to 0 if the previous value was 1 (lines 22 to 25). If the previous value of Q was -1 (line28), we retain the -1.

Line 34 is the "ELSE" that resulted from checking for a positive clock transition. If this point is reached, then there was no valid clock transition nor did a PRESET or CLEAR occur. So Q retains its previous value.

Similarly, we define the function for ~Q as:

```
Function JKFFNOTQ(PRESET, CLEAR, J, K, CLOCK, CLOCKX, NOTQX)
    If PRESET = 0 And CLEAR = 0 Then
        JKFFNOTQ = -1
    Else
    If PRESET = 0 Then
        JKFFNOTQ = 0
    Else
    If CLEAR = 0 Then
        JKFFNOTQ = 1
    Else
    If CLOCK = 1 And CLOCKX = 0 Then
        If J = 0 And K = 0 Then
            JKFFNOTQ = NOTQX
        Else
        If J = 0 And K = 1 Then
            JKFFNOTQ = 1
        Else
        If J = 1 And K = 0 Then
            JKFFNOTQ = 0
        Else
```

86 | Digital Circuit Simulation Using Excel

```
                    If NOTQX <> -1 Then
                        If NOTQX = 0 Then
                            JKFFNOTQ = 1
                        Else
                            JKFFNOTQ = 0
                        End If
                    Else
                        JKFFNOTQ = NOTQX
                    End If
                End If
            End If
        End If
    Else
        JKFFNOTQ = NOTQX
    End If
            End If
        End If
    End If
End Function
```

Two spreadsheets are shown to demonstrate these functions: Figure 4-26 and Figure 4-27. They follow the formula view of the spreadsheet given in Figure 4-25.

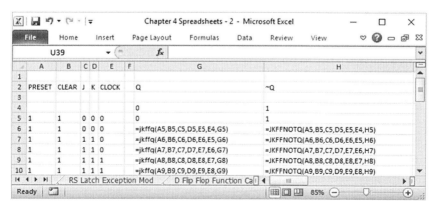

Figure 4-25: Formula View of the Spreadsheet Demonstrating J-K Flip Flop's Response to the J and K Inputs

Figure 4-26 shows the operation of the flip flop to data combinations on J and K. The PRESET and CLEAR inputs are left inactive at a value of 1. Comments are included in Column J to indicate the clocked events.

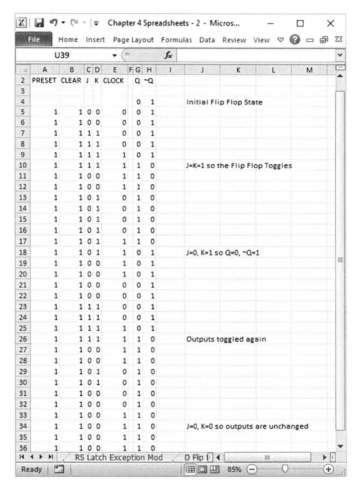

Figure 4-26: Data View of the Spreadsheet Demonstrating J-K Flip Flop's Response to the J and K Inputs

Rows 4 and 5 just start the flip flop off in a known state. After that, the JKFFQ and JKFFNOTQ functions will drive the outputs Q and ~Q in response to J and K.

88 | Digital Circuit Simulation Using Excel

Row 10 shows the flip flop's response to a toggling indication on J and K. Row 18 shows the Q output being set to 0 in response to J=0 and K=1. Then in Row 26 the outputs are again toggled. Finally, Row 34 demonstrates that the outputs do not change when both J and K are 0.

Figure 4-27 executes some scenarios for testing the flip flop's response to conditions related to the PRESET and CLEAR inputs. The formula view of the spreadsheet is not given since it is the same as Figure 4-25 with just a different input pattern.

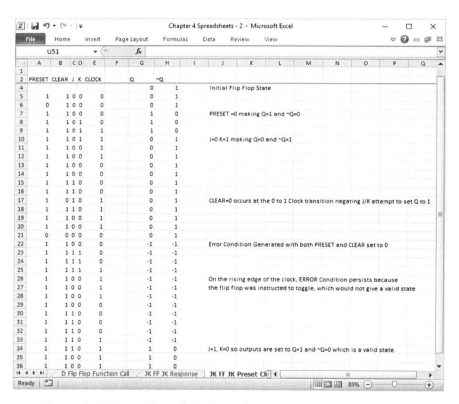

Figure 4-27: Data View of the Spreadsheet Demonstrating J-K Flip Flop's Response to PRESET and CLEAR

Row 6 activates PRESET making Q=1 and ~Q=0 in Row 7. The state is reversed by clocking J=0 and K=1 in Row 9 (output changes in Row 10). CLEAR's interaction with J and K is tested in Row 17. Although J and K try to set Q=1 and ~Q=0, the effect is nullified by the active CLEAR just after the rising edge of the clock. Row 21 introduces the error condition by attempting to assert both PRESET and CLEAR. The rising edge of the CLOCK in Row 25 did not take the flip flop out of the error condition at Row 26 since toggling from the error state would still leave the flip flop in an undetermined, invalid state. So the error persists until Row 34 where J=1 and K=0 sets Q to 1 and ~Q to 0 at the rising CLOCK edge in Row 33.

Further details and operational information for R-S latches and various types of flip flops can be found in References [1], [2], [3], [4], [5], [6], [8], and [9]. Reference [6] also discusses the internal structure of these devices. Reference [3] provides timing diagrams for latches and flip flops.

Shift register design is covered in References [1], [3], [5], [6], and [9].

CHAPTER 5

Asynchronous Design

In this chapter, the Excel simulation tools will be used to analyze an asynchronous circuit. Recall that an asynchronous circuit does not require that all of its clocked devices (e.g. flip flops) operate off of the same timing element. The advantage of doing this is that an asynchronous design is usually simpler than a synchronous one.

The circuit under evaluation is a 4-bit decade counter implemented with J-K flip flops operating in a toggling mode. The goal is for the counting sequence to be

$$0, 1, 2, 3, 4, 5, 6, 7, 8, 9, 0, 1, \ldots$$

Four flip flops are required. They will be configured in an arrangement where the first flip flop, representing the least significant bit, is timed with a free-running clock. Its output is used to clock the next flip flop so that every time the Q output of the first flip flop transitions from 0 to 1, the flip flop representing the next most significant bit is clocked (toggled). This pattern continues for all four flip flops as shown in Figure 5-1.

92 | Digital Circuit Simulation Using Excel

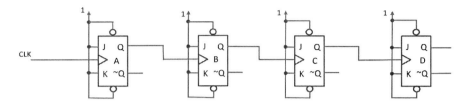

Figure 5-1: J-K Flip Flops Arranged as an Asynchronous Binary Counter

This arrangement will provide a binary count from 0 (0000) to 15 (1111) if the ~Q outputs are used for the count. See Figure 5-2 for the binary count sequence. But for this design, we only want to count to 9 and then have the cycle repeat.

Figure 5-2: Binary Counting Sequence for Asynchronously Clocked J-K Flip Flops

Figure 5-2 shows the clock and the resulting count sequence for the four flip flops. Notice that the way we chose to configure them, the Q outputs actually count down. This isn't a concern, since, as we mentioned, we can just use the ~Q outputs to give us the desired count sequence. **(What would have happened if the flip flops changed state on the negative edge of the clock instead of the positive edge?)**

Next, let's look at a strategy for restarting the counting sequence once it reaches a count of 9. First, we need to add circuitry to detect when a count of 9 is reached. When that occurs, a signal is set that will enable a PRESET pulse to all of the flip flops. The PRESET is used because the ~Q outputs create the count. So to set the ~Q outputs to zero, we can set the Q outputs to 1. At the end of the 9 count's interval, a PRESET pulse will pass to the flip flops and set the ~Q count to zero. Then the PRESET enable will be cleared until the next time that a 9 is detected. A timing diagram showing this sequence is given in Figure 5-3.

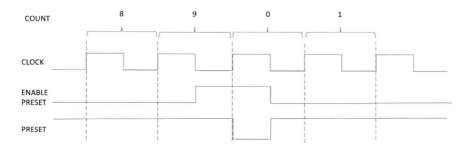

Figure 5-3: Timing Sequence for the Counter's Preset Signal

The PRESET should be generated only after the 9 count has fully completed. It needs to occur immediately at the beginning of the next count. Therefore, a good candidate to use is the first half of the CLOCK signal that makes Flip Flop A increment. However, we only want to have this signal passed to the PRESET of each flip flop at the completion of the 9 count.

94 | Digital Circuit Simulation Using Excel

A viable strategy is to have the preset enabling signal (ENABLE PRESET) go active when the count has reached 9 and the timing CLOCK is low. This occurs during the second half of the 9 count. It will return to the inactive state when the count has gone to 0 and CLOCK is in the second half of that count. This can be easily implemented with an R-S latch.

The PRESET signal itself is derived simply by NANDing the ENABLE PRESET signal with the CLOCK signal.

The complete counter circuit is given in Figure 5-4. The four flip flops are configured in a toggle mode. Except for the first flip flop (designated as "A") which is clocked off of the free running system clock, the others are clocked by the Q output of the previous flip flop. The least significant bit (2^0) is represented by Flip Flop A and the most significant bit (2^3) is represented by Flip Flop D. The R-S latch for generating the PRESET ENABLE is composed of Gates I and J. The 9 count is detected by Gate K and the 0 count by Gate L.

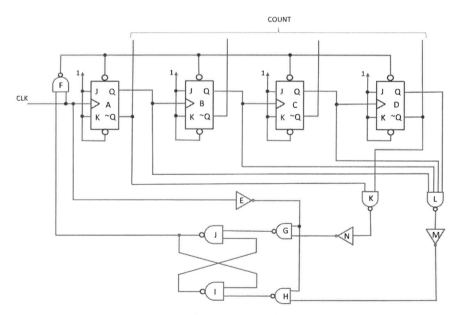

Figure 5-4: Circuit Implementation of the Decade Counter

Chapter 5: *Asynchronous Design* | 95

This circuit is quite easily implemented in the simulation spreadsheet. Since each circuit element, input signal, and other information that we might wish to display is allocated a column, the spreadsheet can become quite wide (i.e. have many columns). For this simulation we will have 22 columns. The spreadsheet (formula view) is shown below in Figure 5-5 to Figure 5-10. Only the first few rows need to be shown since subsequent rows are simply copied down after the first few.

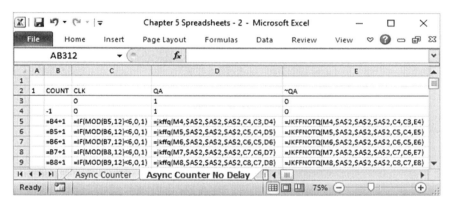

Figure 5-5: Decade Counter Simulation Columns A through E

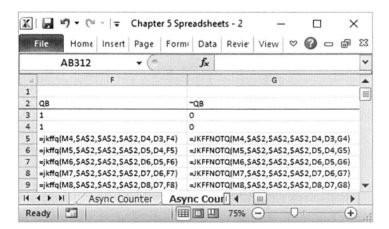

Figure 5-6: Decade Counter Simulation Columns F and G

96 | Digital Circuit Simulation Using Excel

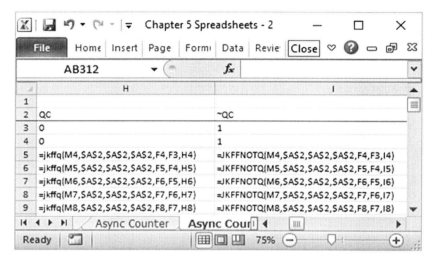

Figure 5-7: Decade Counter Simulation Columns H and I

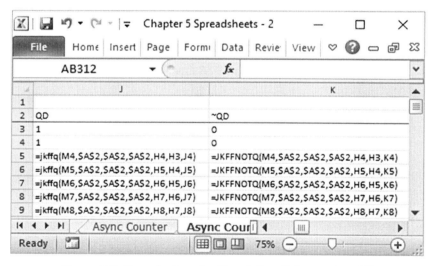

Figure 5-8: Decade Counter Simulation Columns J and K

Chapter 5: *Asynchronous Design* | 97

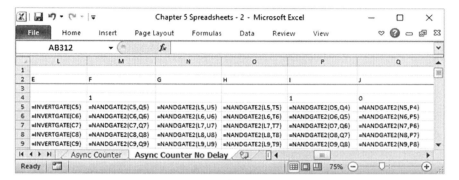

Figure 5-9: Decade Counter Simulation Columns L through Q

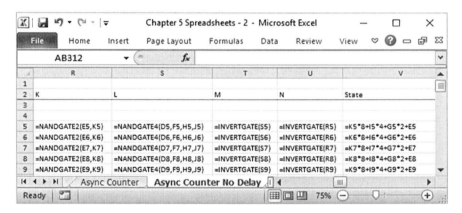

Figure 5-10: Decade Counter Simulation Columns R through V

Let us examine the simulation in detail.

In Figure 5-5, Column A is not used except for Cell A2. It is set to a 1. It is used as the input for unused, active low signals. Essentially, it is the equivalent of a pull-up resistor. So whenever an input signal needs to be set to a constant value of 1, it will reference Cell A2.

Column B is simply a counting reference for each line of the simulation. More importantly, it will also be used to generate the free-running input clock (Column C) for the simulation. The initial value in Cell B4 is set to a -1. Thereafter in subsequent rows, which represent the beginning of the simulation, the count is incremented. The formula entered in Row 5 at the beginning of the simulation is:

$$=B4+1$$

So, Row 5 increments to a 0. The count continues for the rows that follow. Therefore, Row 5 can be copied down for the rows in the rest of the simulation.

Column C simulates the free-running input clock. Rows 3 and 4 have an initial value of 0. This is just to provide valid values for the flip flops to look back two rows. The clock starts running in Row 5. The design has the clock cycle over 12 rows. Six rows are 0 and six rows are 1. As previously mentioned, this is derived from the row count in Column B. For each row number, the remainder from a division by 12 is calculated (this is done by using a "MOD 12" operation). Then the remainder (which takes on the repeating sequence of 0 through 11) is checked to see if it is less than 6. If so, the cell value is set to 0. Otherwise it is set to 1. In other words, for remainders 0 to 5 the clock is 0. For remainders 6 to 11 the clock is 1.

Notice that from Column C onward, Row 2 gives either the names of the signals or the gate reference label from the schematic from which they are generated. Column C contains the label for the generated "CLK" signal. For the columns that represent flip flop states, they have labels of the form "QX" and "~QX." Q implies that this column has the values for the Q output of the flip flop. The X will simply be the label identifier of the flip flop in Figure 5-4. The labels we assigned are A, B, C, and D. The columns with the ~QX header in Row 2 show the ~Q output for Flip Flop X. All but the last of the remaining columns have the label (taken from the schematic) of the simulated gate. The last column ("State") translates the binary count represented by the flip flops to a decimal count.

Returning to Figure 5-5, Column D simulates the Q output of Flip Flop A. Row 2 contains the schematic identifier for the flip flop. Rows 3 and 4 contain an initial value for the Q output. We arbitrarily selected the initial count to be 4, therefore the Q output is a 1. Starting in Row 5, the function representing the Q output of the J-K flip flop is invoked (JKFFQ). The spreadsheet shows the connections made for the 6 input parameters defined for the function. They are:

- PRESET: The schematic shows the PRESET signal as being the output of NAND Gate F. Therefore, it is assigned the value of Column M in the previous row.

- CLEAR: This signal is shown in the schematic connected to a pull-up resistor (i.e. set to a value of 1). Therefore, it references Cell A2. Note that dollar signs are used before the A and the 2. This is so that when the row is copied down, the subsequent rows will have the same absolute reference (actually, A$2 would work as well).

- J: Since the flip flops are operating in the toggle mode, both J and K inputs must be connected to a 1. Therefore the connection is made to Cell A2.

- K: See input J above.

- CLOCK: The schematic has the clock derived from the signal that is being generated in Column C. Since the previous spreadsheet row is used, the reference in Row 5 is set to C4.

- CLOCKX: This signal occurs one spreadsheet row before CLOCK, or two rows back from the current row. Therefore, Row 5 uses the signal in C3.

- QX: This is the value of this parameter taken one row behind. So Row 5 uses D4.

Column E derives the ~Q output for Flip Flop A. Rows 3 and 4 initialize ~QA to a 0. Thereafter it uses the function JKFFNOTQ. The first 6 input parameters are the same as those used in JKFFQ. The only parameter that is different is the seventh one. For this function, the previous value of ~Q is used. So the reference in Row 5 is made to Cell E4.

Figure 5-6 shows Columns F and G. They represent the Q and ~Q states for Flip Flop B. The connections are very similar to Flip Flop A. The PRESET, CLEAR, J, and K inputs are connected exactly the same as Flip Flop A. The differences are the connections to the Clock and the location of the previous state. The clock comes from the Q output of Flip Flop A (Column D). The previous state is taken from the previous row of the same column. For QB, it is Column F and for ~QB, it is Column G. The initial states are Q = 1 and ~Q = 0 as appropriate to give an initial count of 4.

The connections for QC, ~QC, QD, and ~QD (Figure 5-7 and Figure 5-8) follow the same pattern as QB and ~QB. The PRESET, CLEAR, J, and K are the same as for Flip Flops A and B. The CLOCK signals for each come from the Q output of the previous flip flop (consistent with the schematic). The previous states are obtained from the previous row of the same column. The initial states in Rows 3 and 4 give QC = 0, ~QC = 1, QD = 1, and ~QD = 0.

After the flip flops, the next gate in the simulation is inverter E shown in Column L of Figure 5-9. The input to the inverter is the CLK signal. So the entry in Cell E5 is simply:

$$=INVERTGATE(C5)$$

The next entry is for NAND Gate F in Column M. Its two inputs are the signal CLK and the output of NAND Gate J. CLK is the same signal just mentioned as the input for inverter E, so one of the NAND gate inputs will be C5. NAND Gate J's output will be given in Column Q of the spreadsheet in Figure 5-9. So Cell Q5 is picked up as the other input, making the formula in Cell M5:

$$=NANDGATE2(C5,Q5)$$

This gate is also given an initial state in Row 4 since the flip flop simulations beginning in Row 5 require a previous PRESET state. We have it set to the inactive state since the count is 4.

The remaining gates in the circuit are defined through to Column U (Figure 5-10) of the spreadsheet. Their connections are made the same way as the previous two gates just mentioned.

Note that other than the two NAND Gates (I and J) constituting the R-S latch, the signals for all of the other random logic gates do not rely on any previous state. So other than Gates I and J, all of the inputs for the gates in Row 5 are obtained from the outputs of other gates in Row 5. Up to now, we have not needed to account for delays through random logic. Gates I and J, the components of the R-S latch, are initialized so that the output of Gates I and J start off being 1 and 0 respectively.

In Figure 5-10, Column V was added for diagnostic purposes. It converts the count given by the states of the four flip flops into a decimal number. This column is used to see if the count is progressing as designed and if there are any incorrect states. The process is simple. The four flip flops contain a binary representation of the count. To convert to a decimal number:

- Multiply the least significant bit by 2^0 or 1
- Multiply the next most significant bit by 2^1 or 2
- Multiply the next most significant bit by 2^2 or 4
- Multiply the most significant bit by 2^3 or 8
- Add the above four products together.

Finally, Row 5 can be copied down to carry out the simulation as far as desired.

Figure 5-11 to Figure 5-14 show the simulation for one complete decade count. Notice that we arbitrarily started it with a count of 4 because ~QA, ~QB, and ~QD were initialized to 0 while ~QC was initialized to 1.

102 | Digital Circuit Simulation Using Excel

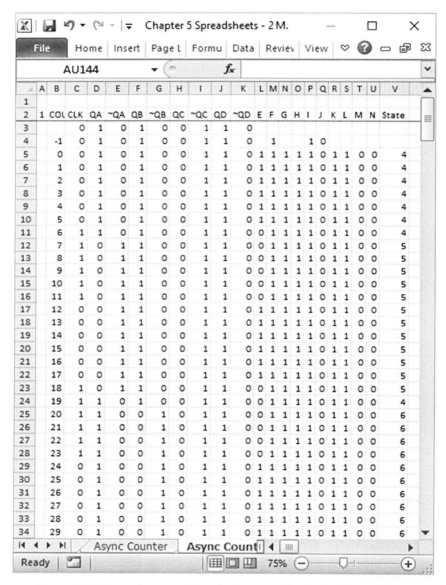

Figure 5-11: Asynchronous Decade Counter Simulation Spreadsheet Rows 1 to 34

Chapter 5: *Asynchronous Design* | 103

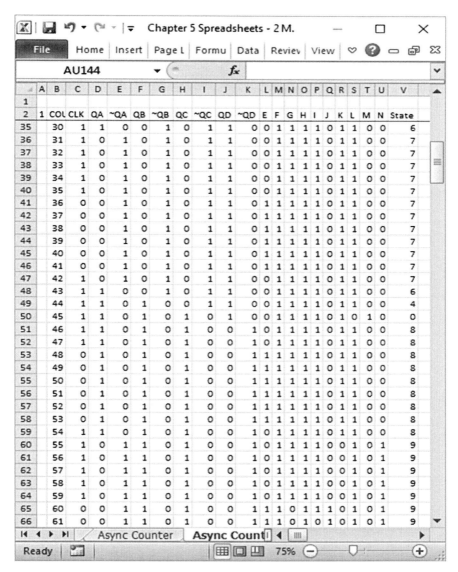

Figure 5-12: Asynchronous Decade Counter Simulation Spreadsheet
Rows 35 to 66

104 | Digital Circuit Simulation Using Excel

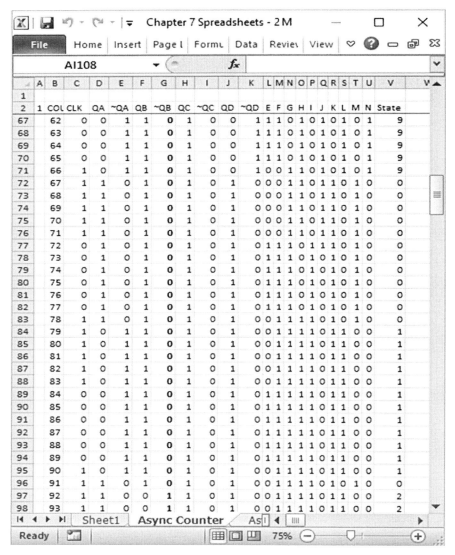

Figure 5-13: Asynchronous Decade Counter Simulation Spreadsheet Rows 67 to 98

Chapter 5: *Asynchronous Design* | 105

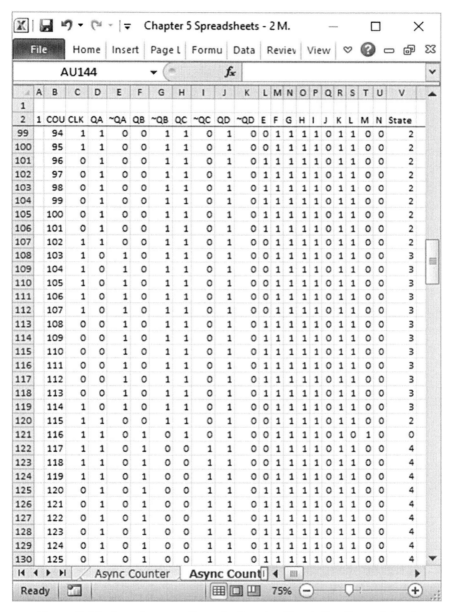

Figure 5-14: Asynchronous Decade Counter Simulation Spreadsheet Rows 99 to 130

Starting with Figure 5-11, Rows 5 through 11 show the count at 4. Then a positive clock transition occurs at Row 11 and Flip Flop A responds to the transition in Row 12. Here the flip flop toggles and QA becomes 0 and ~QA becomes 1. By ~QA becoming a 1, the count changes to a 5 as can be seen in Column V.

In Row 23, we get the next positive transition for CLK which toggles Flip Flop A in Row 24. Here in Row 24, the positive transition of QA, which is the clock to Flip Flop B, toggles the state of QB in Row 25. But notice what happens to the count in Row 24. Flip Flop A changed state but Flip Flop B has not yet responded since the state change lags the clock transition by one row. Therefore the count momentarily becomes 4. Then in Row 25, when Flip Flop B changes state, the count becomes the correct value of 6.

A similar situation happens when the clock undergoes a positive transition in Row 47 in Figure 5-12. The expected transition is from a count of 7 to a count of 8. All of the flip flops are expected to change state, but the changes will ripple through the counter beginning in Row 48. First, flip flop output ~QA goes to a 0, making the count 6. Then in Row 49, ~QB goes to a 0 generating a count of 4. ~QC becoming 0 in Row 50 takes the count to 0. Then finally ~QD transitions to 1, giving the expected count of 8 in Row 51.

These transient counter states are to be expected with an asynchronous circuit. That is why we must be careful where an asynchronous design is used. For example, if we wanted to use the counter to set up a sequence of 10 events, each of which are triggered by the counts that are generated by the circuit, the events triggered by counts of 6, 4, or 0 may be erroneously enabled due to the transient counts at the 7 to 8 transition. If, however, the count was sampled late in the clock cycle, then we would not have this problem. In this example, the clock is in a high state in Rows 47 to 52 and will return to a zero in Rows 53 through 58. If the count is only sampled when the clock is a 0, the count would have completely rippled through the flip flops and, by that time, the value of 8 will be read correctly.

In summary, at each positive edge of CLK the count gets incremented. At those times, we observe:

- Row 11, initiating the count transition from 4 to 5 with no transient counts
- Row 23, initiating the count transition from 5 to 6 with a transient count of 4
- Row 35, initiating the count transition from 6 to 7 with no transient counts
- Row 47, initiating the count transition from 7 to 8 with transient counts of 6, 4, and 0
- Row 59, initiating the count transition from 8 to 9 with no transient counts
- Row 71, initiating the count transition from 9 to 0 with no transient counts (Note that this transition is a bit different in nature than the others and will be examined shortly)
- Row 83, initiating the count transition from 0 to 1 with no transient counts
- Row 95, initiating the count transition from 1 to 2 with a transient count of 0
- Row 107, initiating the count transition from 2 to 3 with no transient counts
- Row 119, initiating the count transition from 3 to 4 with transient counts of 2 and 0

(It is an interesting and a rather simple exercise to examine the transitions of the flip flops to verify that the transient states, and lack of transient states for that matter, are as shown above.)

Looking at where the count reaches 9 (Row 60), some other events occur. In Row 60, NAND Gate K's output goes low since it detected a 9. This causes Inverter N's output to go high which prepares NAND Gate G to be ready to pass the inverted CLK signal that makes the ENABLE PRESET signal active. Since the output of Inverter E is 0, at this point, the output of Gate G remains high. So the R-S latch isn't yet toggled.

At the second half of the 9 count, CLK goes to 0 at line 65. This causes the output of Gate G to go low (also line 65), which puts a 0 to the input of Gate J of the R-S latch. Gate J's output goes high and is fed back to its partner gate in the latch, Gate I. In response, Gate I's output goes to 0 in Row 66. Since the output of the latch at Gate J is 1, Gate F is now ready to pass the PRESET pulse to the flip flops when CLK goes high. This condition happens at Row 71 (Figure 5-13). This makes the output of Gate F go to 0 in Row 71. All of the flip flops respond at Line 72 by setting their ~Q outputs to 0, thereby resetting the count to 0. Also note in Row 71 that the output of Gate G goes back to a 1, which will allow the R-S latch to change state when the output of Gate H goes low.

By having the count go to 0 in Row 72, a few other things happen. First, the output of Gate M, which detects a 0 count, goes high to enable Gate H to change the state of the R-S latch when CLK goes low.

The circuit is now ready to toggle the R-S latch. This happens at Row 77 when CLK goes low in the middle of the 0 count state. The output of Gate E goes high in response to CLK. This drives the output of Gate H to a 0, changing the output of Gate I back to a 1. Once the propagation of I's output through J occurs, the R-S latch has changed state (specifically, Gate J in Row 78) and the enable to Gate F is removed for passing further PRESET signals to the flip flops.

The circuit is back to its normal asynchronous incrementing mode until it reaches the next count of 9.

Chapter 5: *Asynchronous Design* | 109

Looking at the preceding simulation figures, we can see the rippling effect of the count where the output of each flip flop is delayed from its predecessor. This is most apparent on the count transition from 7 to 8. The change in ~QD took 4 rows (Rows 47 to 51) to transition to 1 after the positive edge of CLK occurred. The exception to the rippling is when the count transitions from 9 to 0. Here, the flip flops will all change state at the same time due to the simultaneous application of PRESET. We can see that the transition on ~QD occurs in just one row (Row 71 to 72), which is exactly the same timing as ~QA. Therefore, no transient counts are observed.

Finally, Figure 5-15 shows timing diagrams that are obtained from the spreadsheet. It displays the ~Q outputs for the four flip flops, the PRESET signal used to reset the counter to 0, and the decimal representation of the count. Vertical lines were added to show the transition to each count. The decimal count is indicated just above the ~QA timing diagram.

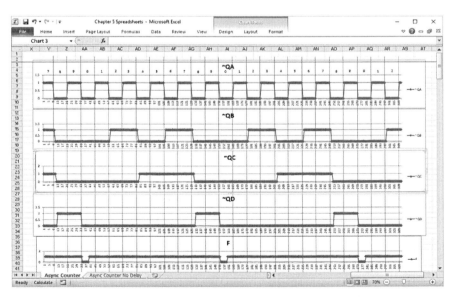

Figure 5-15: Timing Diagram for the Asynchronous Decade Counter

References [1], [5], and [9] review the design of asynchronous counters. References [1] and [5] cover techniques for eliminating transient states.

Reference [9] discusses design considerations for asynchronous sequential circuits.

CHAPTER 6

Synchronous Design

Let us consider a design for a 3-bit synchronous up/down binary counter. This circuit will either count up or down based upon the state of a control line. We will define the control line's function to enable the circuit to count up when it is 1 and down when it is 0. Its label will be "U/~D" shorthand for "UP/~DOWN." There will be three D type flip flops maintaining the count. Flip Flop A will contain the least significant bit, B will have the next most significant bit, and C will have the most significant bit. The Q outputs of each flip flop will define the count.

From this information, we can create a state transition table as shown in Table 6-1.

	U/~D	CURRENT STATE			NEXT STATE		
		QC	QB	QA	QC	QB	QA
COUNT DOWN	0	0	0	0	1	1	1
	0	0	0	1	0	0	0
	0	0	1	0	0	0	1
	0	0	1	1	0	1	0
	0	1	0	0	0	1	1
	0	1	0	1	1	0	0
	0	1	1	0	1	0	1
	0	1	1	1	1	1	0
COUNT UP	1	0	0	0	0	0	1
	1	0	0	1	0	1	0
	1	0	1	0	0	1	1
	1	0	1	1	1	0	0
	1	1	0	0	1	0	1
	1	1	0	1	1	1	0
	1	1	1	0	1	1	1
	1	1	1	1	0	0	0

Table 6-1: State Transition Table for Up/Down Binary Counter

The Current State columns show the state of the Q outputs for each of the three flip flops. The determination for whether the Next State shows the Q outputs incrementing or decrementing their count is based on whether UP/~DOWN is 0 or 1.

Karnaugh maps will be used to convert the state transition table information into a circuit. Three maps need to be created, each one to define the next state of Q for each of the three flip flops. The next states are denoted with an apostrophe (') which is interpreted as a "prime" symbol.

The Karnaugh maps are shown with the prime implicants (not related, or course, with use of the "prime" symbol introduced in the previous paragraph) for QA' in Figure 6-1, QB' in Figure 6-2, and QC' in Figure 6-3.

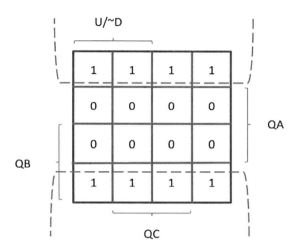

Figure 6-1: Karnaugh Map Defining the Next State of Flip Flop A (i.e. QA')

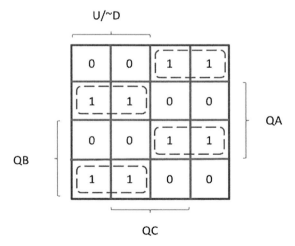

Figure 6-2: Karnaugh Map Defining the Next State of Flip Flop B (i.e. QB')

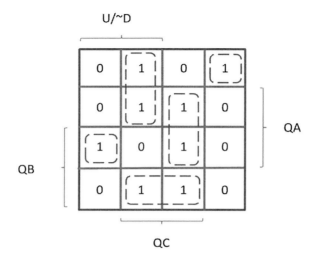

Figure 6-3: Karnaugh Map Defining the Next State
of Flip Flop C (i.e. QC')

The maps give us the following logical sum-of-products equations:

QA' = ~QA
QB' = ~(U/~D) · ~QA · ~QB +
~(U/~D) · QA · QB +
(U/~D) · QA · ~QB +
(U/~D) · ~QA · QB
QC' = (U/~D) · ~QB · QC +
~(U/~D) · QA · QC +
~QA · QB · QC +
(U/~D) · QA · QB · ~QC +
~(U/~D) · ~QA · ~QB · ~QC

These equations define the logical statements presented to the respective D inputs of the flip flops. From these, the circuit can be drawn as shown in Figure 6-4.

Chapter 6: *Synchronous Design* | 115

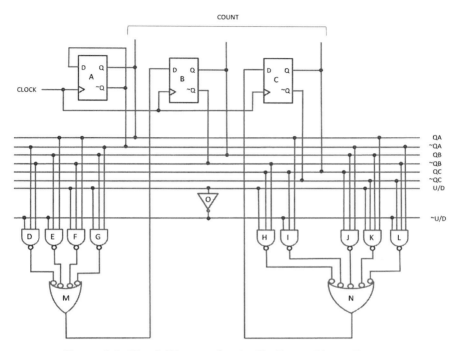

Figure 6-4: Circuit Diagram for the Up/Down Binary Counter

A few more devices need to be added to the library to implement the circuit. Gates D through J are three-input NAND gates. Gates K through M are four-input NAND gates. Gate N is a five-input NAND gate. In standard logic families (e.g. the 74XX/74LSXX family), the following gates are produced:

7410/74LS10 – Three-Input NAND gate

7420/74LS20 – Four-Input NAND gate (already created in Chapter 3)

7430/74LS30 – Eight-Input NAND gate

116 | Digital Circuit Simulation Using Excel

We will follow this convention and add NANDGATE3 and NANDGATE8 to the Excel library.

Note that NANDGATE8 will be used for the five input application for Gate N. Four of the gate's inputs will be connected to the same signal to give the five input functionality.

Because there are 23 columns in the spreadsheet, the formula view is displayed in Figure 6-5 to Figure 6-8.

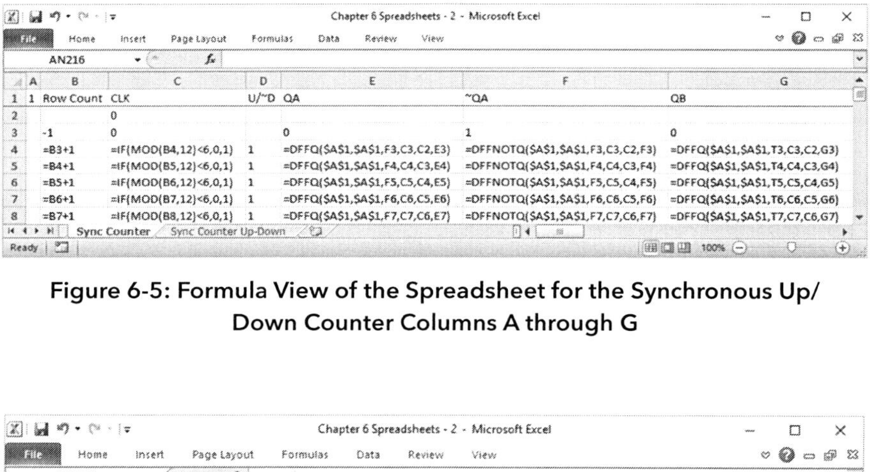

Figure 6-5: Formula View of the Spreadsheet for the Synchronous Up/Down Counter Columns A through G

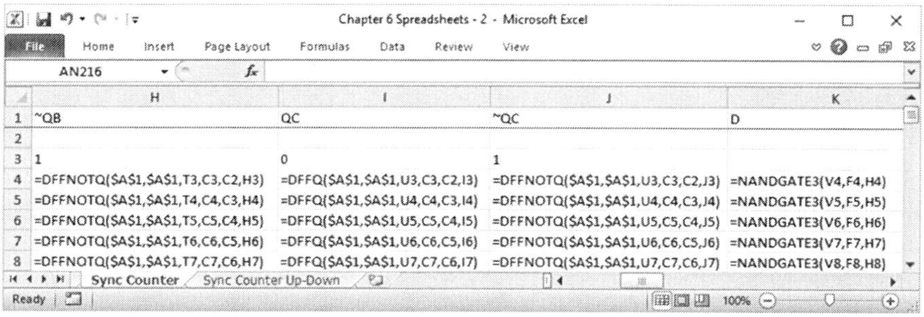

Figure 6-6: Formula View of the Spreadsheet for the Synchronous Up/Down Counter Columns H through K

Chapter 6: *Synchronous Design* | 117

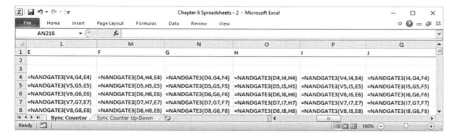

Figure 6-7: Formula View of the Spreadsheet for the Synchronous Up/Down Counter Columns L through Q

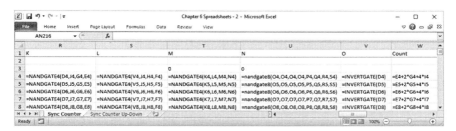

Figure 6-8: Formula View of the Spreadsheet for the Synchronous Up/Down Counter Columns R through W

Column A simply uses Row 1. It provides a logical "1" level for signals which are specified to remain in that state.

Column B contains the row count as we have used before. It is initialized to -1 in Row 3. Each subsequent row increments the value from the previous row. The simulation begins in Row 4.

Column C generates the system clock (CLK) from the row count. It performs a modulo 12 division on the row count so that the clock cycles every 12 rows. A check is performed to see if the remainder from the modulo 12 division is less than 6. If it is less than 6, which will occur in the first half of the cycle, the clock is set to 0. For the second half of the cycle, it is set to 1. Notice that although the simulation and clock generation starts in Row 4, there are two entries of 0 inserted in Rows 2 and 3. This is because the flip flop equations start in Row 4 but need to look back two rows for the clock data. We arbitrarily set them to 0.

Column D defines the up or down counting operation. The value is entered manually depending upon whether we want the circuit to count up (defined by a value of 1 on the signal) or down (defined by a 0).

Starting in Column E and continuing to Column J are the flip flop states. Flip Flop A is defined in Columns E (for QA) and F (for ~QA). Row 3 just sets the initial state. QA is initialized to 0 and ~QA is set to 1. Beginning in Row 4 the D flip flop functions (for Q and ~Q) are called. The first two parameters are not used for this design and are held at a logical 1. They are PRESET and CLEAR. Since they are not used here, they are not shown on the schematic. Continuing in Column E, the next 4 parameters are:

- DATA. This is connected to the previous row of the ~QA output (Column F) as specified in the equation for QA'.

- CLK. The CLK parameter is connected to the system clock represented by the previous row in Column C.

- CLKX. This is the previous clock state with respect to parameter CLK which is also taken from Column C but from the previous row of CLK (i.e. two rows before the current row).

- QX. The last passed parameter is the previous value of Q. It comes from the same column (in this case Column E) but from the previous row.

Column F defines the ~Q output of Flip Flop A. It is similar to Column E. The only difference is that the last parameter (NOTQX) is taken from the previous row in Column F.

The equations for the remaining flip flops (Columns G to J) are analogous to the corresponding columns (Columns E and F) for Flip Flop A. The only major differences are the sources of the data inputs. Let us look at them in detail.

The D input to Flip Flop B comes from the output of NAND Gate M (taken from the previous row in Column T). So that becomes the parameter passed to Flip Flop B's D input for QB (Column G) and ~QB (Column H).

Flip Flop C's data input in Columns I and J comes from the signal generated by Gate N from Column U.

Columns K through U are used to simulate the operation of the NAND (or INVERT-OR) Gates D through N. They have inputs comprised of combinations of QA, ~QA, QB, ~QB, QC, ~QC, U/~D, and ~(U/~D). All of the inputs are taken from the current row of the spreadsheet.

Before looking at the NAND gate simulations, let us temporarily jump ahead to Column V. This is a straightforward application that represents inverter O, which provides the inverted sense of the U/~D signal. Therefore, its input comes from Column D.

Column K simulates the three-input NAND Gate D. From the schematic, its inputs are ~(U/~D), ~QA, and ~QB. These signals are generated in the current row of Columns V, F, and H. Similarly, inputs to the gates represented by Columns L through V are connected to the appropriate cells in much the same way as Gate D. Notice that Columns T and U representing Gates M and N respectively have initial values inserted in Row 3. This is because these gates provide the DATA input to Flip Flops B and C. The simulation starts in Row 4 which requires that data is available in Row 3. For this simulation these values are arbitrarily chosen. They are actually irrelevant until we are near the rising edge of the CLK signal.

Column W gives a decimal representation of the flip flop counts. This operates the same way as the counting function introduced for the

120 | Digital Circuit Simulation Using Excel

asynchronous counter analyzed in the last chapter. As before, it is used as a diagnostic tool to check the validity of the count sequence.

The data view of the spreadsheet is shown in Figure 6-9 to Figure 6-14 for spreadsheet Rows 1 to 217. Note that the row references used in the discussion that follows refer to the spreadsheet index in the left margin. It does not refer to the "Row Count" in Column B.

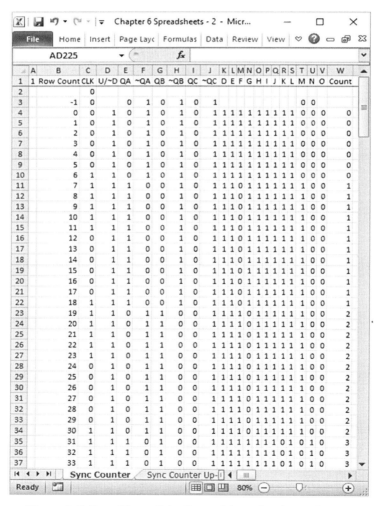

Figure 6-9: Data View of the Spreadsheet for the Up/Down Synchronous Counter Rows 1 to 37

Chapter 6: *Synchronous Design* | 121

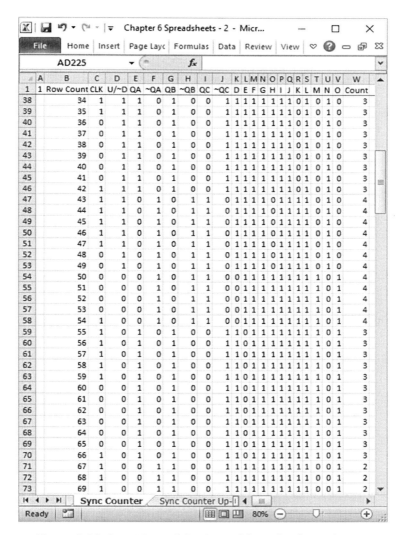

Figure 6-10: Data View of the Spreadsheet for the Up/Down Synchronous Counter Rows 38 to 73

122 | Digital Circuit Simulation Using Excel

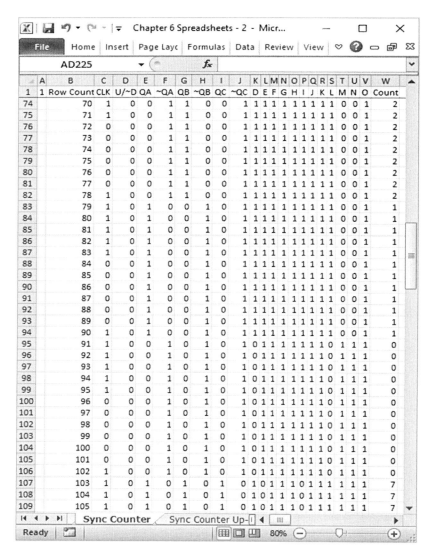

Figure 6-11: Data View of the Spreadsheet for the Up/Down Synchronous Counter Rows 74 to 109

Chapter 6: *Synchronous Design* | 123

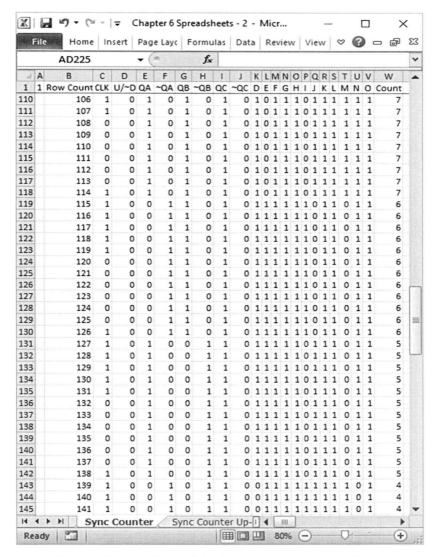

Figure 6-12: Data View of the Spreadsheet for the Up/Down Synchronous Counter Rows 110 to 145

124 | Digital Circuit Simulation Using Excel

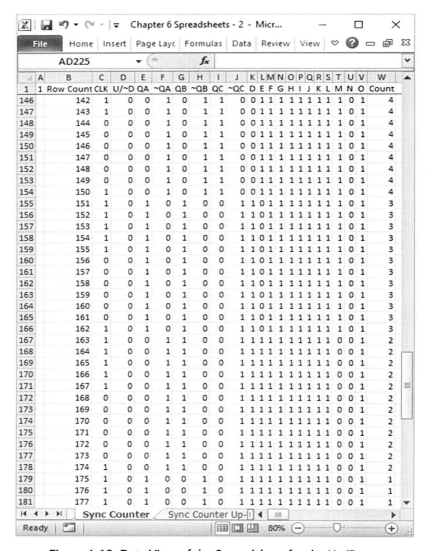

Figure 6-13: Data View of the Spreadsheet for the Up/Down Synchronous Counter Rows 146 to 181

Chapter 6: *Synchronous Design* | 125

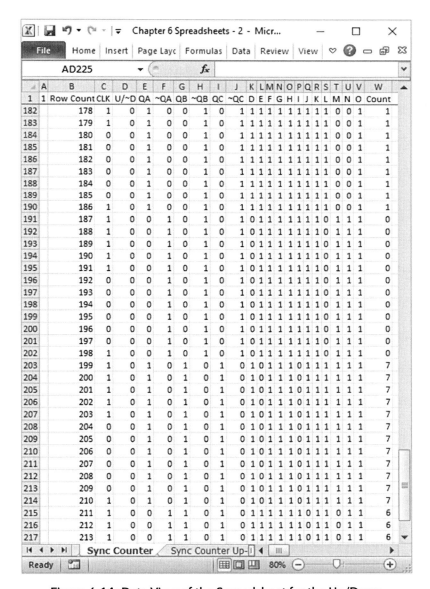

Figure 6-14: Data View of the Spreadsheet for the Up/Down Synchronous Counter Rows 182 to 217

The "UP" counting sequence starts off, since U/~D is a 1, and is run in Rows 4 through 53 (seen in Figure 6-9 and Figure 6-10). Looking at the count in Column W, we can see how the operation of the synchronous circuit differs from the asynchronous counter. Since the clock to all of the flip flops is the same, the count does not have to ripple through all of the devices until it settles. When the count first increments in Row 11, in response to the positive edge of the CLK signal in Row 10, it immediately becomes 1. It stays at that count for 12 rows where it increments to 2 (Row 23). The count of 2 is maintained for 12 rows until the next increment to a count of 3 in Row 35. We find that that the incremented count exactly follows the 12 row cycle of the CLK signal. This can be verified through the last "UP" increment in Row 47.

In Row 54, shown in Figure 6-10, the UP/DOWN control is changed to 0, triggering a down counting sequence. Its operation can be verified by examining Column W. Row 58 (Figure 6-10) shows the first positive clock transition following the instruction to begin down counting. Row 59 shows the count has decremented from 4 to 3. After that the count continues to be decremented by 1 every 12 rows or one clock cycle. Notice also that the 12 row count sequence (Rows 47 to 58) was maintained even when the UP/DOWN control was changed in Row 54.

Timing diagrams for this design are shown in Figure 6-15. The signals shown are:

- UP/DOWN Control Input from Column D
- QA from Column E
- QB from Column G
- QC from Column I
- Decimal value of the binary count from Column W

We can see how the count increments from 0 to 4 until the UP/DOWN signal changes from 1 to 0. Then the count decrements from 4 to 0 and from 7 to 0 thereafter. Again we see evidence of the synchronous switching of all outputs so that no transient counts are generated.

Before leaving the simulation and analysis of this design, it would be useful to look a little further at the UP/DOWN control. This signal was not defined as being synchronized to CLK. Therefore, it may occur anywhere within the period of CLK. Given the ease with which the simulation can be modified, it would be useful to see how the system behaves as the UP/DOWN control is moved with respect to the phase of CLK. Actually, the main concern would be how the circuit behaves as the transition in the control occurs near the rising edge of CLK which is the trigger for the circuit to change state (i.e. increment or decrement).

Figure 6-16 shows the area of the previous simulation (Rows 53 and 54) where the signal U/~D changed from a 1 to a 0. This is where the circuit begins decrementing.

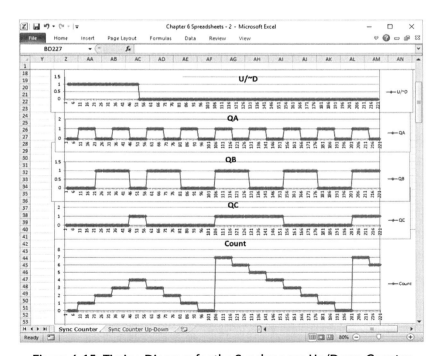

Figure 6-15: Timing Diagram for the Synchronous Up/Down Counter

128 | Digital Circuit Simulation Using Excel

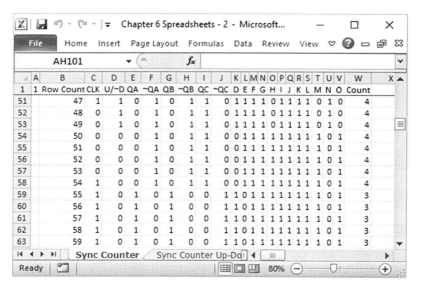

Figure 6-16: Original Spreadsheet Simulation Showing the Transition from Incrementing to Decrementing

By modifying the 1 to 0 transition in Column D and inspecting the updated simulation, the integrity of the counting operation can be tested. In Figure 6-17 to Figure 6-20, the U/~D transition was moved from Rows 55/56 through to Rows 58/59.

The count sequence is unchanged for Figure 6-17, Figure 6-18, and Figure 6-19. The first decremented count from 4 to 3 occurs in Line 59. Then in Figure 6-20 the transition is moved past the rising edge of CLK so the instruction to decrement misses the rising CLK edge occurring at Rows 57/58. Therefore, it continues incrementing and the next count is 5. So the command to decrement is not acted upon until the next CLK rising edge at Rows 69/70. In Row 71 the first decremented count occurs.

Chapter 6: *Synchronous Design* | 129

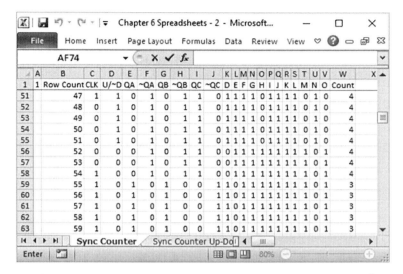

Figure 6-17: Incrementing to Decrementing Transition Moved to Rows 55/56

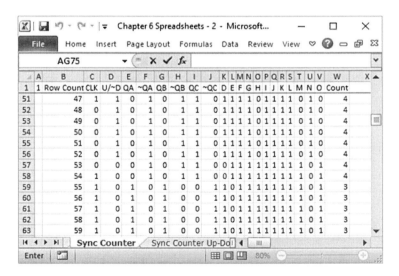

Figure 6-18: Incrementing to Decrementing Transition Moved to Rows 56/57

130 | Digital Circuit Simulation Using Excel

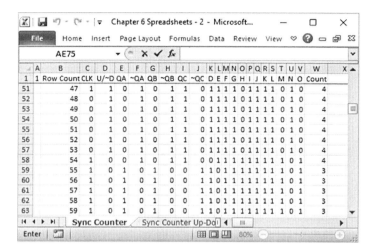

Figure 6-19: Incrementing to Decrementing Transition Moved to Rows 57/58

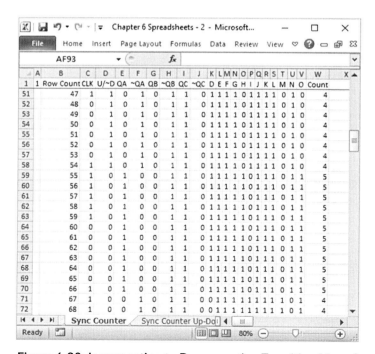

Figure 6-20: Incrementing to Decrementing Transition Moved to Rows 58/59

This specific situation will be revisited in Chapter 7 where propagation delays are discussed. This situation becomes much more involved than is apparent at the present time. We will see how the simulation will uncover the complications that arise when signals undergo delays.

Further information about various types of synchronous counters can be found in References [1], [5], and [6]. Reference [9] has background information about synchronous circuits in general.

Reference [4] covers system timing considerations for sequential circuits. Reference [8] discusses design considerations for determining maximum clock rate for sequential circuits. Reference [2] reviews timing diagrams for counters.

Sequential circuit simplification using Karnaugh Maps is discussed in Reference [6].

Techniques for sequential circuit implementation from state tables (or state diagrams) can be found in References [5] and [9].

CHAPTER

7

Considerations for Propagation Delay and Timing Analysis

Up to this point, we have not directly addressed the question of showing how propagation delays can be simulated. The spreadsheets have been used to verify that the circuit was correctly designed to provide the desired logical output (if a random logic circuit) or sequence (if a sequential circuit).

In this chapter, let us take the next step to see how we can incorporate propagation delays into the simulation and be able to observe their effects. When actual parameters are cited, the 74LS logic family specifications are used. However, as will become clear by the examples, any logic family can be simulated simply by modifying the spreadsheet row from where the inputs are referenced.

NAND Gate

Let us start by considering how to simulate propagation delays through a single random logic gate. The NAND gate is used as an illustrative example. The technique can be directly applied to any other gate.

First we will review the NAND gate with no delay, which is the situation we have considered up to now. Figure 7-1 shows the familiar formula view of the spreadsheet. Here, it is assumed that the inputs shown in the spreadsheet, Rows 4 to 27, represent a sequence in time (note that the simulation actually starts in Row 7). Also, for deriving each output, the inputs are taken from the same row. Let us also assume that that each row of the spreadsheet represents the same fixed time interval, say, for example, 10 ns (which is approximately the propagation delay through a 74LS00 NAND gate).

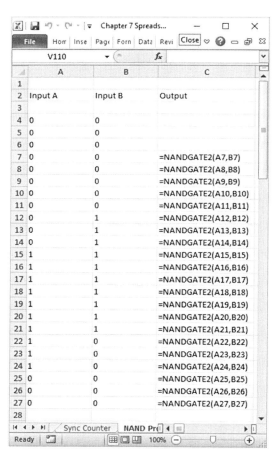

Figure 7-1: Formula View for the Spreadsheet Showing a NAND Gate with No Propagation Delay

Chapter 7: Considerations for Propagation Delay and Timing Analysis | 135

Figure 7-2 has the spreadsheet's data view which also serves as a truth table. Timing diagrams are also given, which are generated from the truth table. The vertical lines show the interval when both inputs are 1, producing a low on the output.

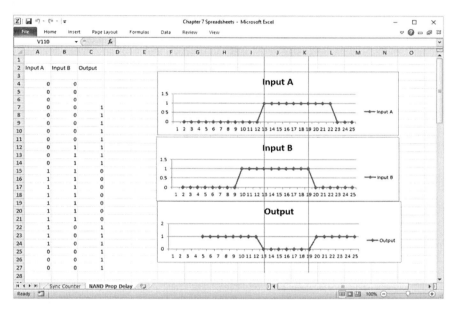

Figure 7-2: Data View and Timing Diagrams for the Spreadsheet Showing a NAND Gate with No Propagation Delay

Now we want to model a NAND gate where there is a propagation delay which is equivalent to one row of the spreadsheet (10 ns). In order to do this, the inputs used for a given output are simply taken from the previous row. See Figure 7-3. The data view of the spreadsheet with the accompanying timing diagrams show the result (Figure 7-4). We can see that the period over which the output is low is delayed by one interval or one spreadsheet row. That is, both inputs are 1 for one row (Row 47) before the output goes low. The output remains low for one row (Row 54) after Input B has gone low.

136 | Digital Circuit Simulation Using Excel

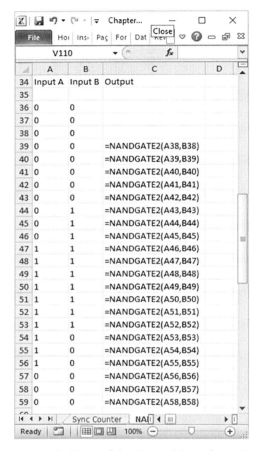

Figure 7-3: Formula View of the Spreadsheet for a NAND Gate with a One-Row Delay

Chapter 7: *Considerations for Propagation Delay and Timing Analysis* | 137

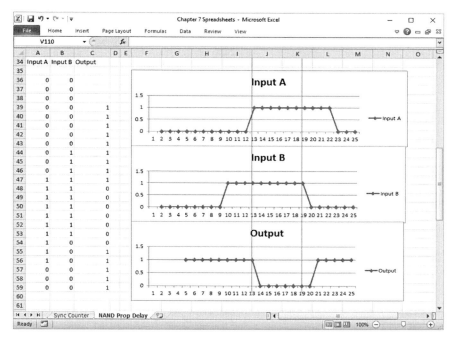

Figure 7-4: Data View of the Spreadsheet for a NAND Gate with a One-Row Delay

This can be carried out for additional delays by adjusting the row providing the input reference. Figure 7-5 and Figure 7-6 show a delay of two rows which would be the equivalent of a 20 ns propagation delay.

138 | Digital Circuit Simulation Using Excel

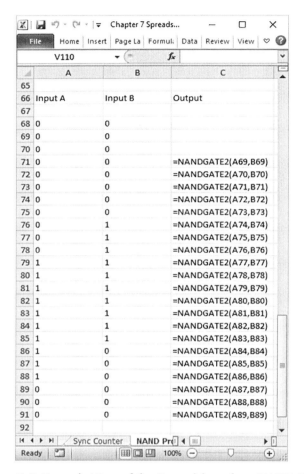

Figure 7-5: Formula View of the Spreadsheet for a NAND Gate with a Two-Row Delay

Chapter 7: *Considerations for Propagation Delay and Timing Analysis* | 139

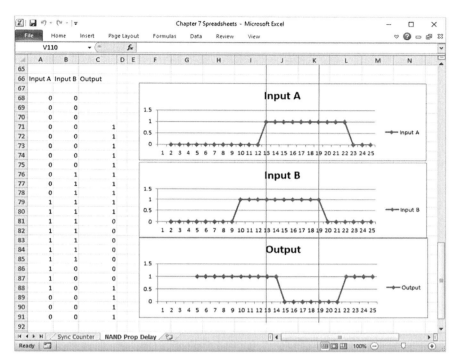

Figure 7-6: Data View of the Spreadsheet for a NAND Gate with a Two-Row Delay

The timing diagrams in Figure 7-6 show the output delayed by 20 ns with respect to Figure 7-2.

This strategy can be directly applied to simulate propagation delays on inverters and other types of simple gates (see Chapter 1).

Combinatorial Logic

In the simulation covered in Chapter 3 for combinatorial logic, a new spreadsheet row was created for each unique combination of inputs. This allowed us to verify that the truth table or equations were correctly realized by the circuit. However, it does not look at how the circuit reacts dynamically to changing input signals. Therefore, it is considered a "static" verification of the design. That is, while that spreadsheet is a valuable tool that is used to ensure that the circuit correctly implements the desired truth table, it does not address the question of how the physical components with real propagation delays will respond to changing input signals. For example, transient responses due to possible race conditions and switching hazards are not analyzed.

By representing each input combination with just one row in the spreadsheet, we assumed that the input to output transitions occur "instantaneously." So the output shown is the steady state result of the inputs after all of the gates in the circuit have been allowed to "settle out" and the output has become stable.

Let us consider at how we can look at the dynamic behavior of a simple random logic circuit. By "dynamic," we are referring to its response to a sequence of inputs changing with time and subject to propagation delays through the circuit.

Consider the simple circuit shown in Figure 7-7.

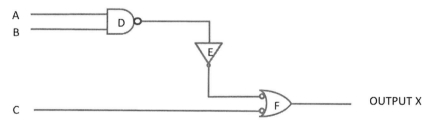

Figure 7-7: Example for Illustrating Propagation Delays

Normally, propagation delays through most of the commonly used logic families are quite small. For the LS family of gates, the delay is typically in the order of 10 nanoseconds (ns) up to 20 ns for the maximum delay. In sequential circuits, specifically synchronous circuits, cumulative propagation delays should be insignificant when compared to system clocks.

Let us first consider inverter E. Using the approach defined up to now without propagation delay considerations, when we look at its called function in a spreadsheet located at, let us say, Row 9 and Column E with its input taken from Column D, it would be:

=Invertgate(D9)

It says, "If the input at Column D Row 9 (the current row) is 0, then output a 1; otherwise output a 0."

In order to introduce a propagation delay of one row, all that needs to change is the cell referenced for the input. Instead of being the current row, it would be the previous row. Therefore the Row 9 entry would change to:

=Invertgate(D8)

It says, "If the input located at Column D Row 8 (the previous row) is 0, then output a 1; otherwise output a 0." Therefore, the propagation delay for the gate is represented by one row of the spreadsheet.

The same thing can be done to the NAND gates. The function call in Column D Row 9 for a one-row propagation delay with inputs from Columns A and B would be:

=NANDGATE2(A8,B8)

It says, "If the inputs located at Row 8 for Columns A and B are both 1, output a 0; otherwise output a 1."

Let us first simulate the circuit in Figure 7-7 without accounting for propagation delay. The formula view, using function calls, is given in Figure 7-8. The data view for the spreadsheet in Figure 7-8 is given in Figure 7-9.

Let us also interpret the inputs as time-varying signals as we proceed down the spreadsheet. We will take the duration of each row as 10 ns which is approximately the typical propagation delay for the 74LS00

(NAND) and 74LS04 (Inverter) gates. The input switching is shown to be no faster than 80 ns or 8 rows.

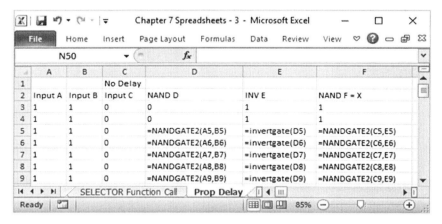

Figure 7-8: Analysis with no Propagation Delay Considerations – Formula View

Chapter 7: *Considerations for Propagation Delay and Timing Analysis* | 143

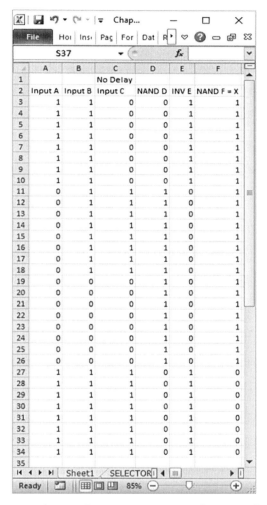

Figure 7-9: Analysis with no Propagation Delay Considerations – Data View

Here, we consider the input columns to be showing how Input A, Input B, and Input C are changing with time. Each row is showing the transition of each signal from its previous row. So it is describing a sequence in time. Since each row represents a time interval of 10

144 | Digital Circuit Simulation Using Excel

ns, Figure 7-9 with its 32 rows (Rows 3 through 34) represents a total time duration of 320 ns.

Given the input pattern, the output appears as expected. Outputs change immediately in response to inputs. A set of timing diagrams based on Figure 7-9 are given in Figure 7-10. It operates as expected. That is, it follows the truth table that would be generated by the static interpretation of the circuit components.

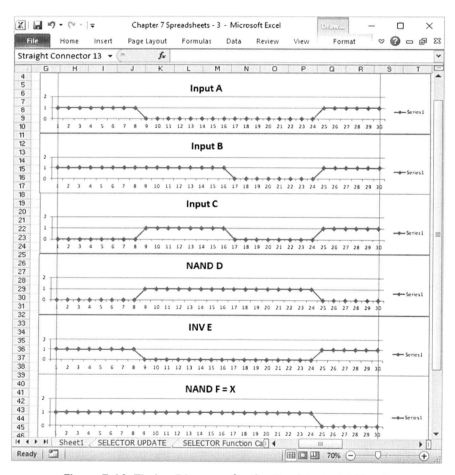

Figure 7-10: Timing Diagrams for the Combinatorial Circuit with no Delay

Chapter 7: *Considerations for Propagation Delay and Timing Analysis* | 145

Now the spreadsheet is updated for both the formula and data views of the circuit (Figure 7-11 and Figure 7-12) allowing for propagation delays of one spreadsheet row for all gates. The associated timing diagrams are given in Figure 7-13.

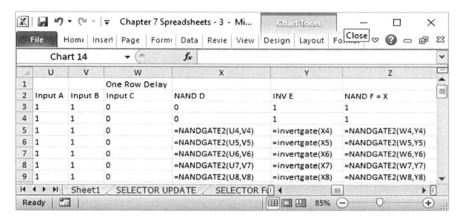

Figure 7-11: Analysis for the Combinatorial Circuit with One-Row Propagation Delay – Formula View

146 | Digital Circuit Simulation Using Excel

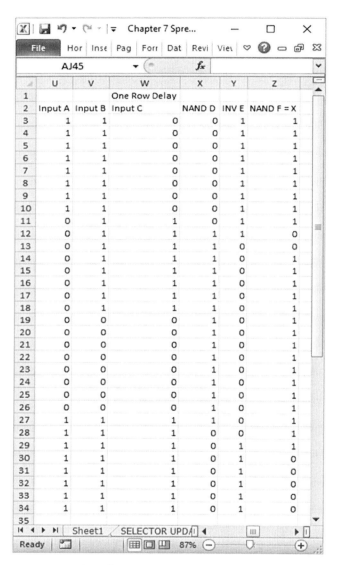

Figure 7-12: Analysis for the Combinatorial Circuit with One-Row Propagation Delay – Data View

Chapter 7: *Considerations for Propagation Delay and Timing Analysis* | 147

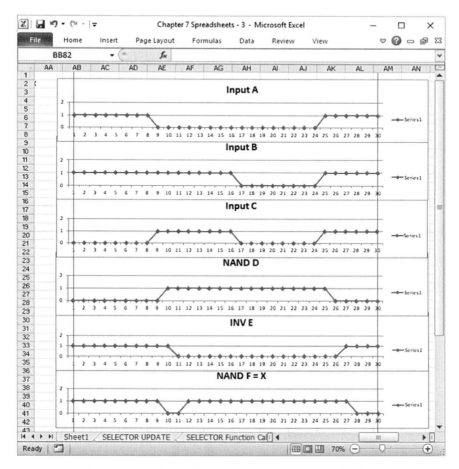

Figure 7-13: Timing Diagrams for Combinatorial Circuit with One-Row Delay

There are several points to notice. In the original results derived with no propagation delay (Figure 7-9), Rows 3 through 26 output a 1 as expected by the formulas. Figure 7-12, allowing for propagation delays, has 1 being output in Rows 3 through 29 with a brief interval (in Rows 12 and 13) outputting 0. The extension of the "1" output sequence beyond Row 26 is due to the fact that changes in the input required the additional delay (3 rows representing 3 gates) to affect the output. The spurious 0's at Rows 12 and 13 are due to a switching hazard caused by unequal delays through the circuitry of two inputs that changed simultaneously. In Row 11, inputs A and C changed at the same time. Logically, the output should have remained the same both before and after the transition. However, input C has a shorter path to the final gate and arrived there before the change in input A. So for that brief period of time, both inputs to the final NAND (or INVERT OR) gate are 1 making the output go to 0.

This situation is not uncommon. It is expected that outputs need time to settle. In critical cases, where the transient outputs cannot be tolerated, the output is usually sampled at some point beyond which the worst case delay path would have already been settled at the output.

Also note that in Figure 7-9, the output changed to the stable value of 0 in Row 27, since it responded to inputs in the same row. Figure 7-12, on the other hand, shows the same stabilization happening in Row 30, which is due to the three-gate propagation delay.

Figure 7-14 (formula view) and Figure 7-15 (data view) extend the delay once more by simulating a two row (or 20 ns) delay which would represent the maximum delay time for all gates.

	AP	AQ	AR	AS	AT	AU
1			Two Row Delay			
2	Input A	Input B	Input C	NAND D	INV E	NAND F = X
3	1	1	0	0	1	1
4	1	1	0	0	1	1
5	1	1	0	=NANDGATE2(AP3,AQ3)	=invertgate(AS3)	=NANDGATE2(AR3,AT3)
6	1	1	0	=NANDGATE2(AP4,AQ4)	=invertgate(AS4)	=NANDGATE2(AR4,AT4)
7	1	1	0	=NANDGATE2(AP5,AQ5)	=invertgate(AS5)	=NANDGATE2(AR5,AT5)
8	1	1	0	=NANDGATE2(AP6,AQ6)	=invertgate(AS6)	=NANDGATE2(AR6,AT6)
9	1	1	0	=NANDGATE2(AP7,AQ7)	=invertgate(AS7)	=NANDGATE2(AR7,AT7)

Figure 7-14: Formula View of Combinatorial Logic with a Two-Row Delay

Chapter 7: *Considerations for Propagation Delay and Timing Analysis* | 149

In the data view given in Figure 7-15, the circuit behaved as expected for the additional delay. The transient 0 counts were doubled to four rows (Rows 13 to 16) and the stable 0 output in response to all 1's at the input was delayed another three rows, from Row 30 in Figure 7-12 to Row 33 in Figure 7-15.

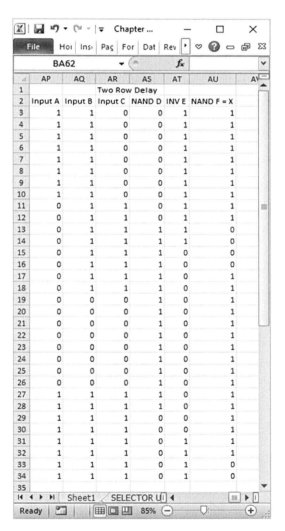

Figure 7-15: Data View of Combinatorial Logic with a Two-Row Delay

Flip Flops

Chapter 4 defined the appropriate spreadsheet rows from which the various flip flop inputs (e.g. CLEAR, CLOCK, DATA, etc.) should be taken based on the device's expected operation. As we have done with simple gates, to simulate propagation delay in flip flops, all that needs to be done is to take the appropriate inputs from an earlier row. It is important to note that when we identify the previous value of Q or (~Q) that is used to retain the state when there is no change due to PRESET, CLEAR, or clocking, it must always come from the immediately preceding row.

It is very easy to demonstrate how this works. Consider the spreadsheet shown in Figure 7-16. It will use the example of a D flip flop.

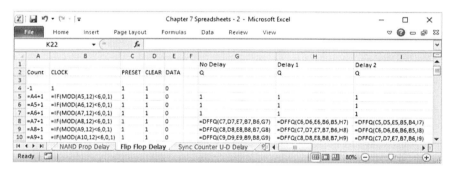

Figure 7-16: Formula View of Spreadsheet Providing Various Delays for a D Flip Flop

Column A provides the row count as has been used before.

Column B will generate the clock to the flip flop using modulo 12 division, giving the clock a period of 12 rows.

Columns C, D, and E give values for the PRESET, CLEAR, and DATA inputs respectively. The pattern was manually added to demonstrate the delays.

Chapter 7: *Considerations for Propagation Delay and Timing Analysis* | 151

Column G shows the flip flop behavior with no added delay. This is the model that has been used up to now. Notice that the simulation does not start until Row 8. This is simply to allow for some initialization.

After three rows of initialization, Column H introduces a delay of one additional spreadsheet row starting in Row 8. Notice that the corresponding parameters are simply taken one row earlier than those in Column G. The exception, of course, is the last parameter representing the previous state of Q.

Column I adds one more row of delay than Column H. But again, the previous state of Q is taken from the previous row. The data view of the spreadsheet is given in Figure 7-17 and Figure 7-18.

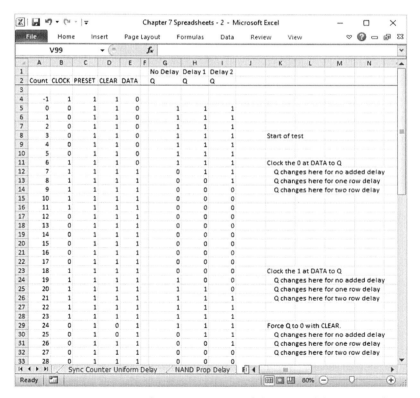

Figure 7-17: Data View for Rows 1 to 33 of the Spreadsheet Providing Various Delays for a D Flip Flop

152 | Digital Circuit Simulation Using Excel

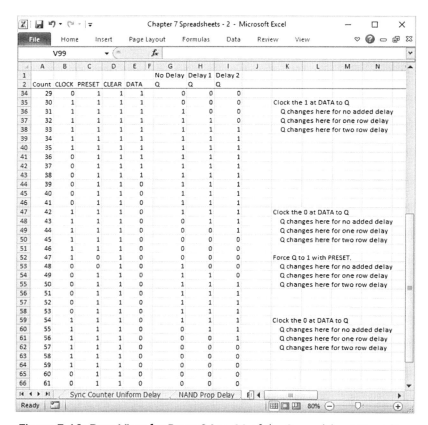

Figure 7-18: Data View for Rows 34 to 66 of the Spreadsheet Providing Various Delays for a D Flip Flop

The first event that triggers a state change occurs in Excel Row 11. There is a positive transition of the clock while DATA is 0. This should make the Q output 0. In Column G, with no added delay, it happens in Row 12. In Column H with one-row delay, it occurs in Row 13. Column I's two-row delay shows it happening in Row 14. The same delay is seen where a 1 on DATA gets clocked to Q in Rows 24 to 26.

Row 29 shows Q being set to 0 by activating CLEAR in Rows 29 and 30. The same delays as before are observed for the three situations in Rows 30 to 32.

In Figure 7-18, the same delays are demonstrated for PRESET activation (Row 52) as well as for some additional data clocking.

It should also be pointed out that different propagation delays can be set for each of the following

- PRESET to Q
- CLEAR to Q
- DATA clocked to Q

if necessary.

Asynchronous Counter

It would be illustrative, at this point, to revisit the designs from the previous chapters to see how accounting for propagation delay affects them.

Let us first consider the asynchronous decade counter analyzed in Chapter 5. At that time, it was assumed that the signal propagation through the discrete gates occurs within the time duration of one iteration, or row, of the spreadsheet. Now let us see how introducing delays through the discrete gates and flip flops affects the performance of the design.

Looking back at the schematic in Figure 5-4, a delay of one spreadsheet row will be added to Gates E through N. Note that this also includes the latch comprised of Gates I and J, which is in addition to the delay through the partner gate that was originally introduced to ensure proper operation (see Chapter 4).

An additional delay is also added to the Q and ~Q outputs of all the flip flops. In order to do this and maintain the same row numbering as Figure 5-5 through Figure 5-10, the column headings in Row 2 are moved up to Row 1 and some additional initialization numbers are added in Row 2 since the additional delay requires that the flip flops reference data back to Row 2. Note that no initialization is required for CLK Cell C2. Since Cell C3 is a 0, no positive clock transition can occur. No change is made to the COUNT or CLK signal in order to give us the same row reference for comparisons between spreadsheets. The modified formula view of the spreadsheet is given in Figure 7-19 for Columns A through E of the spreadsheet. It includes the modifications for Flip Flop A. The remaining flip flops (not shown) undergo analogous changes. Modifications for Gates E through I (Columns L through P) are shown in Figure 7-20. Figure 7-21 has the modifications for Gates J through N (Columns Q through U).

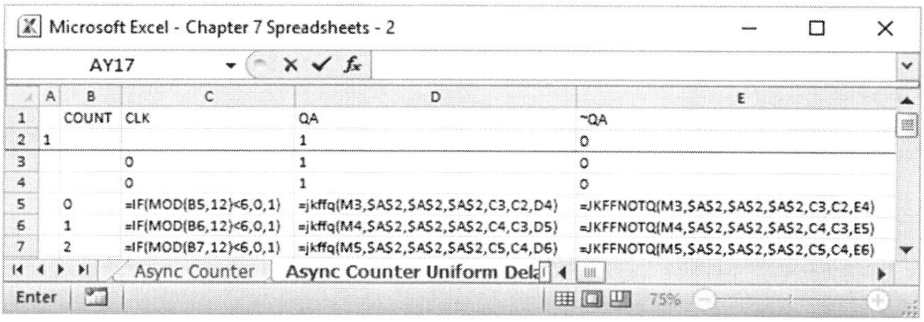
Microsoft Excel - Chapter 7 Spreadsheets - 2
AY17

	A	B	C	D	E
1		COUNT	CLK	QA	~QA
2	1			1	0
3			0	1	0
4			0	1	0
5		0	=IF(MOD(B5,12)<6,0,1)	=jkffq(M3,A2,A2,A2,C3,C2,D4)	=JKFFNOTQ(M3,A2,A2,A2,C3,C2,E4)
6		1	=IF(MOD(B6,12)<6,0,1)	=jkffq(M4,A2,A2,A2,C4,C3,D5)	=JKFFNOTQ(M4,A2,A2,A2,C4,C3,E5)
7		2	=IF(MOD(B7,12)<6,0,1)	=jkffq(M5,A2,A2,A2,C5,C4,D6)	=JKFFNOTQ(M5,A2,A2,A2,C5,C4,E6)

Async Counter | Async Counter Uniform Del
Enter 75%

Figure 7-19: Formula View for Columns A thorough E of the Spreadsheet for the Asynchronous Counter with One-Row Delay

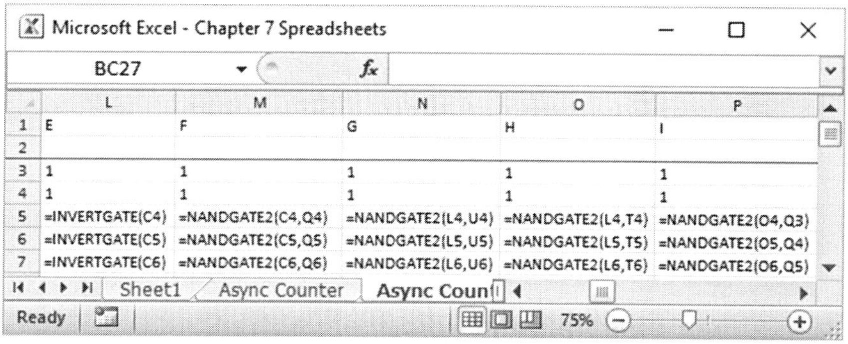
Microsoft Excel - Chapter 7 Spreadsheets
BC27

	L	M	N	O	P
1	E	F	G	H	I
2					
3	1	1	1	1	1
4	1	1	1	1	1
5	=INVERTGATE(C4)	=NANDGATE2(C4,Q4)	=NANDGATE2(L4,U4)	=NANDGATE2(L4,T4)	=NANDGATE2(O4,Q3)
6	=INVERTGATE(C5)	=NANDGATE2(C5,Q5)	=NANDGATE2(L5,U5)	=NANDGATE2(L5,T5)	=NANDGATE2(O5,Q4)
7	=INVERTGATE(C6)	=NANDGATE2(C6,Q6)	=NANDGATE2(L6,U6)	=NANDGATE2(L6,T6)	=NANDGATE2(O6,Q5)

Sheet1 | Async Counter | Async Coun
Ready 75%

Figure 7-20: Formula View for Columns L thorough P of the Spreadsheet for the Asynchronous Counter with One-Row Delay

Chapter 7: *Considerations for Propagation Delay and Timing Analysis* | 155

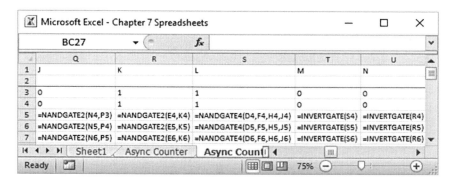

Figure 7-21: Formula View for Columns Q thorough U of the Spreadsheet for the Asynchronous Counter with One-Row Delay

Figure 7-22 gives a side-by-side comparison for the simulations of the counter for the "no delay" and "one-row delay" configurations.

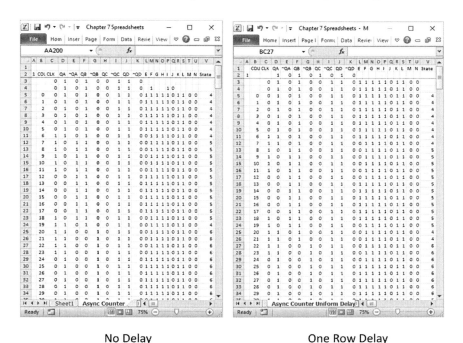

No Delay   One Row Delay

Figure 7-22: No Delay and One-Row Delay Comparison of Rows 1 to 34 of the Asynchronous Counter Simulation

In both cases, the simulation starts in Row 5. The first positive clock transition occurs between Rows 10 and 11. With no delay, the count increments via Flip Flop A in Row 12. With the one-row delay, the count increments in Row 13.

The next event happens where there is the positive clock transition that occurs at Rows 22 and 23. The count is expected to change from 5 to 6. With no delay, we see that Flip Flop A toggles, making the least significant bit change from a 1 to a 0, giving a transitory count of 4 at Row 24. The positive transition on QA is presented on the CLOCK input of Flip Flop B, causing it to change its output in Row 25, which gives the expected count of 6.

Looking at the case where there is a one-row delay, Flip Flop A changes state at Row 25 (i.e. one row later than 24, as expected). But then there is the additional delay through Flip Flop B. So the transient count of 4 persists for two rows; 25 and 26. The eventual 6 count is reached at Row 27.

This is the type of behavior we would expect when adding delays. So far there is no problem if we followed the guidelines mentioned in the previous chapter. There we noted that if we sample the count in the second half of the clock (Rows 29 through 34), there should be no problem misinterpreting the transient count as a valid state.

Let's look further along in the simulation comparison in Figure 7-23. It looks at Rows 34 to 63 in the spreadsheets. The next increment happens where CLK transitions in Rows 34 and 35. Only the least significant bit of the count will change (Flip Flop A). So there is a clean transition from 6 to 7 in Row 36 when there is no delay. With the one-row delay, there is still a clean transition, but it happens one row later in Row 37.

The next CLK transition occurs at Rows 46 and 47. Here we expect the count to change from 7 to 8. As noted before, there are some transient states as the count ripples through all four flip flops. With no delay, the sequence of counts becomes:

- Row 48 for transient count 6
- Row 49 for transient count 4
- Row 50 for transient count 0
- Row 51 for stable count 8

With the one-row delay, CLK still transitions at Rows 46 and 47, but as before when the count transitioned from 5 to 6, each transient count persists for an extra row as the signals propagate through the flip flops. So now the count sequence becomes:

- Rows 49 and 50 for transient count 6
- Rows 51 and 52 for transient count 4
- Rows 53 and 54 for transient count 0
- Row 55 for stable count 8

Notice that it took four more rows to reach the stable count. With the additional delay, the transient count of 0 extends into the second half period of CLK, which begins at Row 53. Earlier, we proposed that sampling the counts during the second half of the clock cycle would ensure that only valid counts would be read. Clearly that would no longer be true with the delays shown here. At this point, let us examine the simulation in more detail to see the other consequences of having a transient 0 count occur in the second half of the clock cycle.

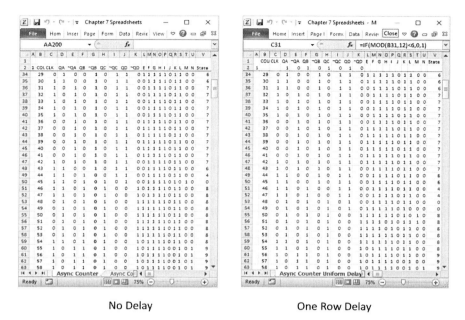

Figure 7-23: No Delay and One-Row Delay Comparison of Rows 34 to 63 of the Asynchronous Counter Simulation

Recall that Gate H is used to detect when the count reached 0 (via Gates L and M) while CLK was low (via Gate E) to reset the latch comprised of Gates I and J. Looking at Rows 56 and 57 in the one-row delay case of Figure 7-23 (Column O), Gate H is creating a resetting pulse that is presented to the latch. This did not happen when no delay was simulated (see the "No Delay" spreadsheet in Figure 7-23). Gate H output a 1 during this time as originally designed. Let us see how this happened.

Gate H uses the outputs of Gates E and M as inputs. The extraneous reset pulse results from the detection of a 0 count from the flip flops through Gate L and then Gate M. In this case, it is the transient count of 0 in Rows 53 and 54 that was detected. Figure 7-24 shows what happens in both the delayed and non-delayed cases when the spurious 0 count is detected by Gate M. It shows the signals that are combined via NAND Gate H to produce its output.

Chapter 7: *Considerations for Propagation Delay and Timing Analysis* | 159

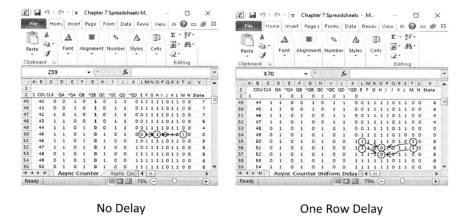

No Delay   One Row Delay

Figure 7-24: Consequences Resulting from the Transient 0 Count

In the no delay situation, the transient 0 state occurs while the signal from Gate E is still low in Row 50. This is the time during which CLK is high. As a result, the output of Gate H remains high, so no pulses are generated.

When there is the one-row delay, there are two places where Gate H produces a 0 output. These occur in Rows 56 and 57. This differs from the no delay case by virtue of the fact that now when the output of Gate M goes high, it has been delayed to when CLK is low (specifically, to when the output of Gate E is high). Again, this is due to propagation delays through the flip flops.

As it turns out, this particular situation does not create a problem with the circuit's functioning. By generating this extra pulse to the latch via Gate H, it is only going to drive it into the state in which it already exists (i.e. the output of Gate I is already 1). There would be a problem if there were a transient 9 count at this particular time. Then it would cause the count to reset prematurely by setting the output of Gate J to a 1 enabling PRESET to the flip flops. As we saw earlier, a transient 9 count is not possible. This situation does point out the usefulness of this tool by simulating these various timing situations.

Let us continue with the analysis of this circuit to look at another interesting area. This is the simulation around where the flip flops are preset to a value of 0. This is shown in Figure 7-25.

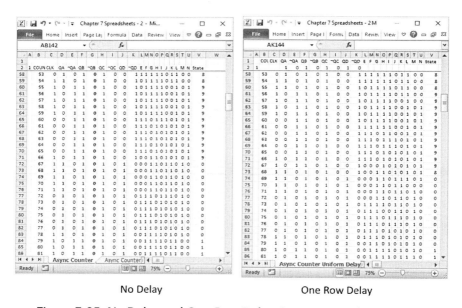

No Delay   One Row Delay

Figure 7-25: No Delay and One-Row Delay Comparison of Rows 58 to 86 of the Asynchronous Counter Simulation

The process begins when a count of 9 is detected by Gate K. This happens in Row 60 with no delay and 62 with the one-row delay. The logical sequence has already been described in Chapter 5 and is very straightforward. One significant consequential event is to see when the latch comprised of Gates I and J enables the PRESET. This occurs when the output of Gate J goes high. With no delay, this happens at Row 65 and the delayed circuit has it occurring at Row 68. The important point to note is that the PRESET ENABLE happens

Chapter 7: *Considerations for Propagation Delay and Timing Analysis* | 161

when CLK is still low. The PRESET itself is generated when CLK goes high.

The 0 to 1 transition of CLK happens at Rows 70 and 71. It produces the PRESET when the output of Gate F goes low. With no delay, this happens in Row 71 and there is a direct transition from a 9 to a 0 count in Row 72. With the delay, the PRESET occurs in Row 72 and there is a momentary count of 8 (Row 73) in the 9 to 0 transition. This is due to the positive CLK transition (Rows 70 and 71) toggling Flip Flop A one row before the PRESET could affect it and all of the other flip flops.

Because the PRESET is presented to all the flip flops at the same time, the circuit does not experience rippling counts. There is just a uniform delay due to the same propagation delay from PRESET to Q and ~Q through each flip flop.

One can also experiment with the spreadsheet to see the ramifications of additional delays or delays in another area. Let us consider the area just discussed where the flip flops are PRESET. It would be interesting to see what might happen if additional delays were present. We start with the current spreadsheet for the one-row delay. In order to keep these tests simple, only the delay on Gate F will be manipulated.

First, consider Gate F to have a two-row propagation delay. The formula at Row 65 becomes:

$$=NANDGATE2(C63,Q63).$$

The data view of the resulting spreadsheet is shown in Figure 7-26.

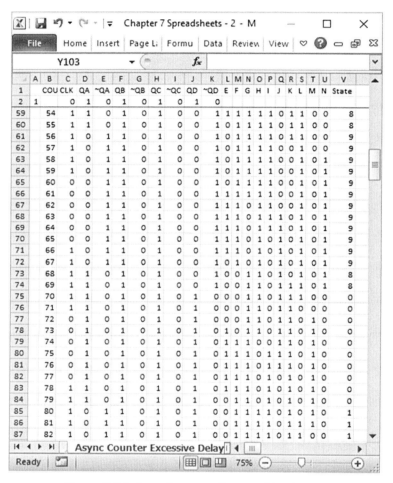

Figure 7-26: Asynchronous Counter Simulation
with a Two-Row Delay on PRESET

One difference has occurred. At Row 74, there is an additional transient count of 8 as compared to Figure 7-25. This occurred because the PRESET is now delayed two rows after the rising edge of CLK toggled Flip Flop A.

Chapter 7: *Considerations for Propagation Delay and Timing Analysis* | 163

One more iteration is shown in Figure 7-27. Here, the delay for Gate F is 4 Rows. So the formula for Gate F in Row 65 is:

=NANDGATE2(C61,Q61)

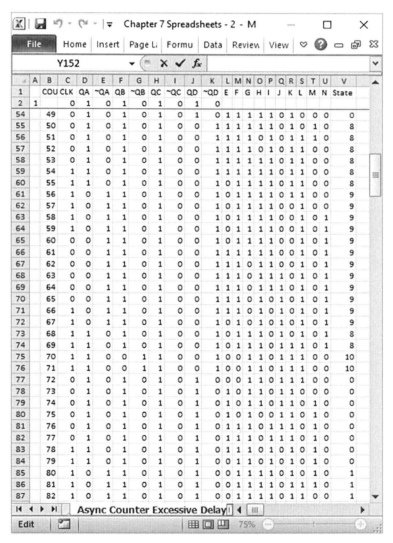

Figure 7-27: Asynchronous Counter Simulation with a Four-Row Delay on PRESET

This situation is even more interesting and illustrates the potential timing issues these simulation techniques can uncover. In addition to seeing the transient 8 count, there is also a transient count of 10 occurring in Rows 75 and 76. Here, the PRESET delay allowed enough time for the count to ripple through Flip Flop A to Flip Flop B which changed the 8 count to a 10 before the PRESET could affect the flip flops. Figure 7-28 gives the waveform showing the counting sequence.

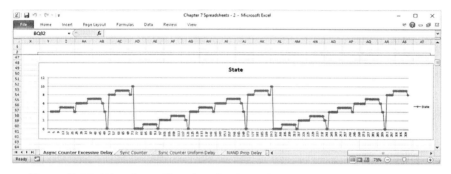

Figure 7-28: Waveform Showing the Asynchronous Counting Sequence with Excessive PRESET Delay

Synchronous Counter

In this section, the synchronous counter described in Chapter 6 is examined to see how it reacts to propagation delays and other timing considerations.

An additional single-row delay is incorporated into the spreadsheet simulation as shown in Figure 7-29 through Figure 7-32. These can be compared to Figure 6-5 through Figure 6-8 in Chapter 6, which shows the spreadsheet without the additional delay.

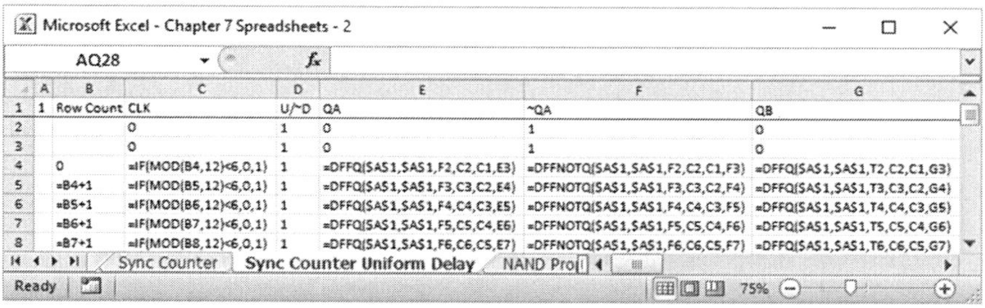

Chapter 7: *Considerations for Propagation Delay and Timing Analysis* | 165

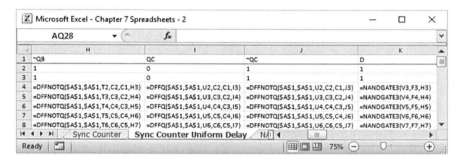

Figure 7-29: Formula View of the Synchronous Up/Down Counter Spreadsheet Columns A through G with One-Row Delay

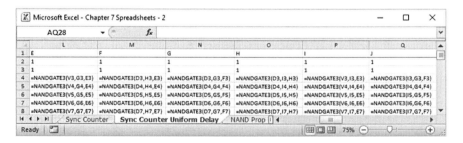

Figure 7-30: Formula View of the Synchronous Up/Down Counter Spreadsheet Columns H through K with One-Row Delay

Figure 7-31: Formula View of the Synchronous Up/Down Counter Spreadsheet Columns L through Q with One-Row Delay

166 | Digital Circuit Simulation Using Excel

Figure 7-32: Formula View of the Synchronous Up/Down Counter Spreadsheet Columns R through W with One-Row Delay

A side-by-side comparison of the data view of the non-delayed and delayed simulation is shown in Figure 7-33 to Figure 7-35 for Rows 1 through 109. Note that in Rows 1 through 53, the count is incrementing, while Row 54 onward has the counter decrementing. This is controlled by the U/~D input which, interestingly, is an asynchronous signal in the synchronous application. This is will be looked at more closely later on.

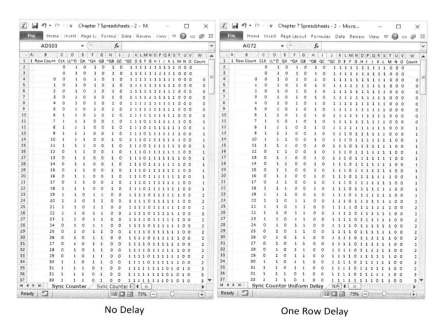

No Delay   One Row Delay

Figure 7-33: No Delay and One-Row Delay Comparison of Rows 1 to 37 of the Synchronous Counter Simulation

Chapter 7: *Considerations for Propagation Delay and Timing Analysis* | 167

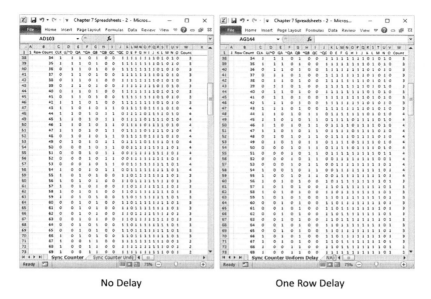

No Delay   One Row Delay

Figure 7-34: No Delay and One-Row Delay Comparison of Rows 38 to 73 of the Synchronous Counter Simulation

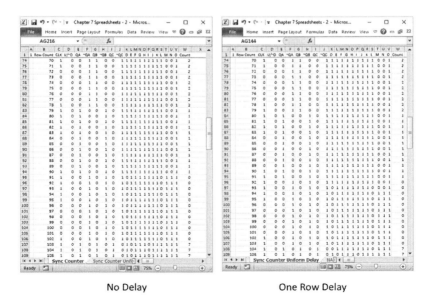

No Delay   One Row Delay

Figure 7-35: No Delay and One-Row Delay Comparison of Rows 74 to 109 of the Synchronous Counter Simulation

The most interesting point to note is that except for the added delay of one row, the counter sequences are identical. This is mainly the result of using a synchronous design. The rippling count situation does not occur. The flip flop outputs, generating the count, all change simultaneously. In this situation, as long as the components are fast enough to allow the signals to set up at the D inputs to the flip flops before the next rising edge of CLK, the counting sequence will perform as expected.

Looking at Figure 7-33, the one-row propagation delay through Flip Flop A is seen where its first state change is in Row 11 without a delay and in Row 12 with the delay. This signal is then propagated through Gate F in Column M. Here we see the additional delay through Gate F delays its output from Row 11 to Row 13. When the output of Gate M finally changes from 1 to 0 to set up the D input of Flip Flop B, we see that it picked up an additional delay of 3 rows (i.e. Row 11 vs Row 14). However, this signal still gets to the D input of Flip Flop B before the next rising edge of CLK which occurs in Row 22. As long as the cumulative delay is less than 11 rows (the number of rows between rising CLK edges) the circuit will operate correctly.

As mentioned in Chapter 6, the U/~D control that is used to control the counting operation is not synchronized to CLK. With no considerations for propagation delay, it was found that the location of the U/~D transition within a cycle of CLK had no effect on the counting sequence integrity. That is, while there may be a delay in reacting to the U/~D signal, there were no unexpected jumps in the sequence. It would be interesting to see if the situation changes due to the introduction of propagation delays.

Figure 7-36, Figure 7-37 and Figure 7-38 show the 1 (increment operation) to 0 (decrement operation) transition of the U/~D signal moved to Rows 54/55, 55/56, and 56/57 respectively.

Chapter 7: *Considerations for Propagation Delay and Timing Analysis* | 169

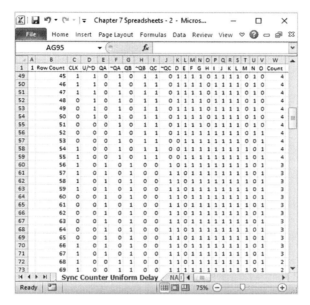

Figure 7-36: Synchronous Counter Up/Down Control Transition at Rows 54/55

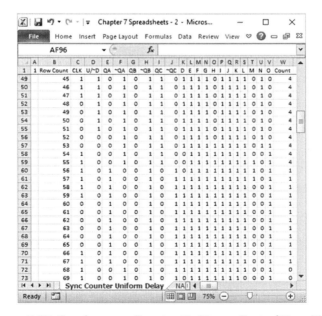

Figure 7-37: Synchronous Counter Up/Down Control Transition at Rows 55/56

170 | Digital Circuit Simulation Using Excel

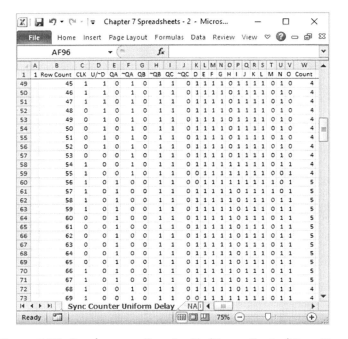

Figure 7-38: Synchronous Counter Up/Down Control Transition at Rows 56/57

In Figure 7-36 there is no change in the circuit's operation from the right side of Figure 7-34 where the U/~D transition occurred at Rows 53/54. The first decrement from 4 to 3 occurs in Row 60 for both situations.

Figure 7-37 shows some unexpected behavior. The count transitioned from 4 to 1. It did not either increment or decrement. We note that after assuming the count of 1, on the next positive transition of CLK, the proper decrementing operation begins. Row 72 shows the count decrementing from 1 to 0. From that point on, it continues the decrement correctly. Before examining what is happening here, let us look at what happens in the next figure.

In Figure 7-38, the U/~D command transition moves to Rows 56/57. At that point, it is too late to change the operation of the circuit from

Chapter 7: *Considerations for Propagation Delay and Timing Analysis* | 171

incrementing to decrementing. Gates M and N are not updated in time to produce a count of 3 on the CLK transition in Rows 57 and 58. So the count continues to increment from 4 to 5. However, in Row 72 the decrementing process begins. Essentially, the decrementing process was delayed by a count. Figure 7-39 shows the waveforms resulting from this situation. Notice that count incremented once more after the U/~D control changed state. This is simply due to the propagation delay keeping the D inputs of Flip Flops B and C from reacting to the changed U/~D signal in time for the next 0 to 1 CLK transition.

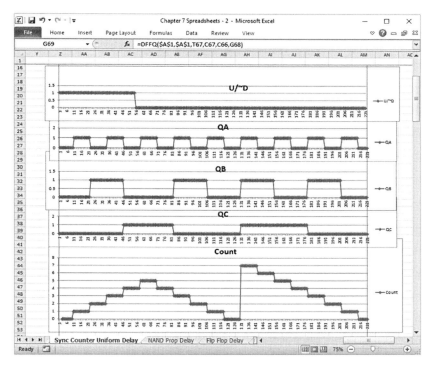

Figure 7-39: Waveforms Showing Delayed Response to U/~D

At this point, let us use the simulation to see what happened in Figure 7-37 to cause the counter to jump from 4 to 1 rather than decrement correctly from 4 to 3.

In order for the count to decrement to a 3, the flip flop states must become QA=1, QB=1, and QC=0. But the states actually became QA=1, QB=0, and QC=0. It is clear that the CLK transition at Rows 57/58 changed Flip Flop A's state correctly so that QA transitioned to a 1 in Row 60. Flip Flop C also changed state to a 0 in Row 60 as expected. So let us proceed with the assumption that Flip Flop B did not change state properly. It should have transitioned to a 1 rather than stay at 0.

To analyze what has happened, Figure 7-40 focuses on the spreadsheet in the area of the problem. Cell G60 (bordered with a dark square in the figure) was expected to become 1. The signals which control the flip flop state at G60 are circled. The other input signals that are held at a logical 1 are not shown.

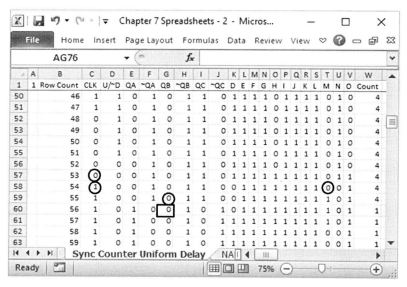

Figure 7-40: Examination of QB and its Controlling Inputs

The positive transition of CLK can be seen occurring in Cells B57 and B58. The previous state is in G59. The value for the D input is in

Chapter 7: *Considerations for Propagation Delay and Timing Analysis* | 173

Cell T58. From the spreadsheet, it is clear that T58 is a 0, which is the reason that QB did not change to a 1.

Column T defines the output of NAND Gate M. This gate controls the D input of Flip Flop B and should be a 1 to make QB a 1 upon the next clock transition. The gate has four inputs, which are provided by Gates D, E, F, and G. This is shown in Figure 7-41. Comparing the circled inputs in Row 57 with the expected operation in Figure 7-36, it is clear that the output of NAND Gate D did not become a 0 as before in Cell K57. Figure 7-42 looks at the inputs controlling the value of NAND Gate D at Cell K57. We can see that the problem is that the output of Inverter O in Cell V56 is not a 1 as it was in Cell V56 of Figure 7-36. The input driving Inverter O is the U/~D signal. This is the signal that has been delayed under the tests that have just been executed. So we can conclude that moving the transition of U/~D to Cells 55/56 provided enough delay to the updating of the D input to Flip Flop B to miss the CLK transition in Rows 57/58.

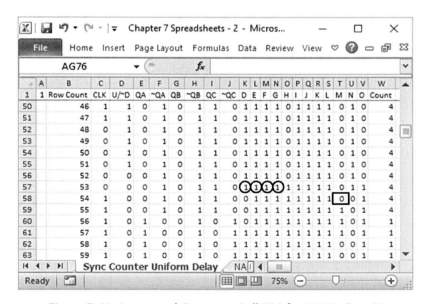

Figure 7-41: Inputs and Output at Cell T58 for NAND Gate M

174 | Digital Circuit Simulation Using Excel

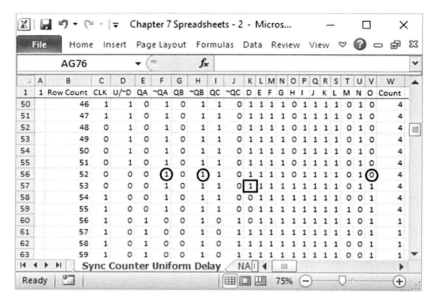

Figure 7-42: Inputs and Output at Cell K57 for NAND Gate D

It is clear that the delay in U/~D would not have an effect on Flip Flop A since the value for its D input is directly derived from its ~Q output, which only experiences the propagation delay through the flip flop. But it would be instructive to see why there was no problem with Flip Flop C under this scenario which also used U/~D to derive its D input.

The same analysis as before is carried out for Flip Flop C (see Figure 7-43). Its output correctly changed to a 0 in Cell I60. This is because the signal to its D input in Cell U58 (NAND Gate N) changed to a 0 in time. But also notice that in the preceding row it was a 1.

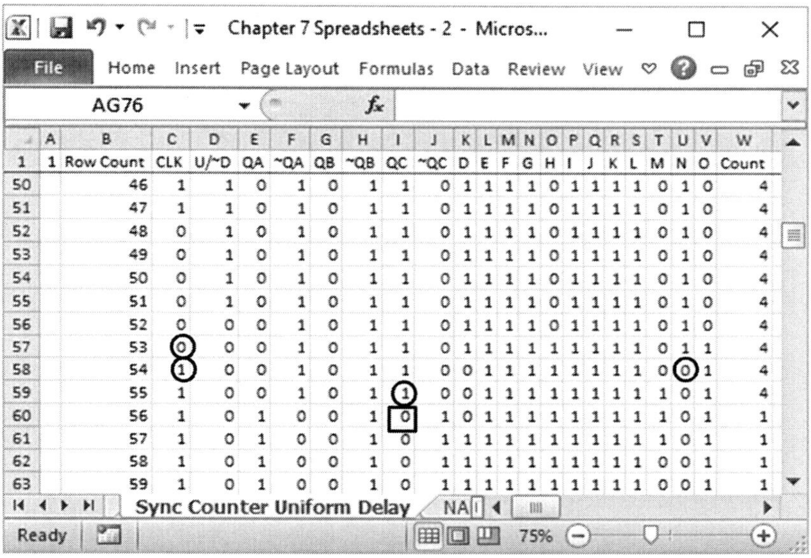

Chapter 7: *Considerations for Propagation Delay and Timing Analysis* | 175

Row	Row Count	CLK	U/~D	QA	~QA	QB	~QB	QC	~QC	D	E	F	G	H	I	J	K	L	M	N	O	Count
50	46	1	1	0	1	0	1	1	0	1	1	1	1	0	1	1	1	1	0	1	0	4
51	47	1	1	0	1	0	1	1	0	1	1	1	1	0	1	1	1	1	0	1	0	4
52	48	0	1	0	1	0	1	1	0	1	1	1	1	0	1	1	1	1	0	1	0	4
53	49	0	1	0	1	0	1	1	0	1	1	1	1	0	1	1	1	1	0	1	0	4
54	50	0	1	0	1	0	1	1	0	1	1	1	1	0	1	1	1	1	0	1	0	4
55	51	0	1	0	1	0	1	1	0	1	1	1	1	0	1	1	1	1	0	1	0	4
56	52	0	0	0	1	0	1	1	0	1	1	1	1	0	1	1	1	1	0	1	0	4
57	53	0	0	0	1	0	1	1	0	1	1	1	1	1	1	1	1	1	0	1	1	4
58	54	1	0	0	1	0	1	1	0	0	1	1	1	1	1	1	1	1	0	0	1	4
59	55	1	0	0	1	0	1	1	0	0	1	1	1	1	1	1	1	1	1	0	1	4
60	56	1	0	1	0	0	1	0	1	0	1	1	1	1	1	1	1	1	1	0	1	1
61	57	1	0	1	0	0	1	0	1	1	1	1	1	1	1	1	1	1	1	0	1	1
62	58	1	0	1	0	0	1	0	1	1	1	1	1	1	1	1	1	1	0	0	1	1
63	59	1	0	1	0	0	1	0	1	1	1	1	1	1	1	1	1	1	0	0	1	1

Figure 7-43: Examination of QC and its Controlling Inputs

The inputs and output for Gate N are given in Figure 7-44. All of the inputs are 1 as they should be. The key input to look at is Gate H in Column O. It is the signal that just changed in time to make the output of Gate N become 0, unlike the situation that occurred for Gate M with respect to Flip Flop B. The inputs to Gate H are analyzed in Figure 7-45.

176 | Digital Circuit Simulation Using Excel

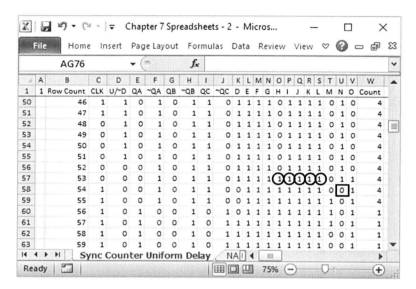

Figure 7-44: Inputs and Output at Cell U58 for NAND Gate N

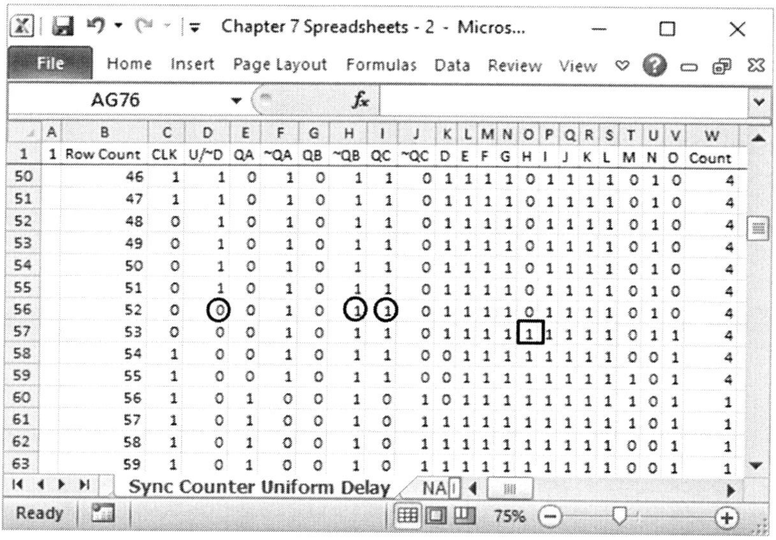

Figure 7-45: Inputs and Output at Cell O57 for NAND Gate H

Chapter 7: *Considerations for Propagation Delay and Timing Analysis* | 177

In the figure, it is clear that the output of Gate H became a 1 because the input from Column D (representing the U/~D signal) is a 0. Notice that this is exactly the point where the U/~D signal just becomes a zero.

It is now clear what happened. The signal coming from Gate N did reach Flip Flop C in time to change its output from a 1 to 0 in response to the decrement directive coming from U/~D. The U/~D signal that had just changed was able to affect the D input of Flip Flop C before the next rising edge of CLK. If it had reached the gate any later, it would not have reached Flip Flop C in time. The reason that the correct signal from Gate M did not reach Flip Flop B in time is that it needed to use the inversion of the U/~D signal. The additional delay through inverter O delayed the signal to the D input of Flip Flop B by just enough time that it missed the rising edge of CLK.

Figure 7-46 shows the timing waveforms that are generated by the spreadsheet. The discontinuity in the timing becomes apparent on the "Count" waveform immediately following where the U/~D signal transitions to a 0.

Simulations such as these provide useful information to the circuit designer. It might not matter if the response to the U/~D command was delayed until the next cycle as long as the increment/decrement sequence was correct. But the simulation shows that if different sections of the circuit react to U/~D at different times, it can cause the count sequence to be in error. The designers may need to consider options to synchronize U/~D with the system clock. For example, the original U/~D signal may be retimed by latching it into a D flip flop on the rising edge of CLK so that the Q output of this new flip flop becomes the U/~D control to the rest of the circuit.

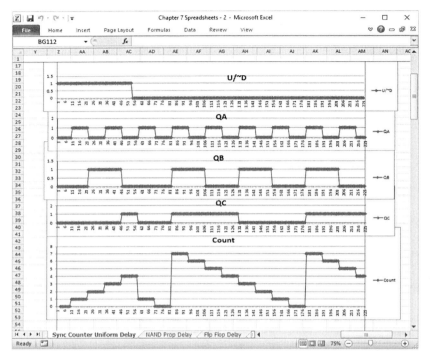

Figure 7-46: Waveforms Showing the Counting Error Due to U/~D Timing

Reference [3] provides some introductory information about propagation delay.

Reference [4] discusses propagation delay in combinatorial circuits. It also looks at how to synchronize asynchronous signals to the clock of a synchronous circuit.

CHAPTER
8

Variable Delay Analysis

Let us look a little further into how the Excel simulations allow us to examine circuits whose components interact with variable delays.

Consider the following simple circuit in Figure 8-1.

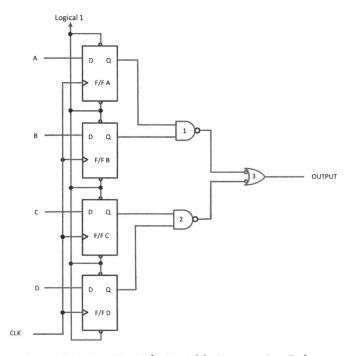

Figure 8-1: Test Circuit for Variable Propagation Delays

180 | Digital Circuit Simulation Using Excel

It consists of four D-type flip flops labeled F/F A, F/F B, F/F C, and F/F D. The D input of each flip flop connects to signals A, B, C, or D (the designated signal connects to the flip flop with the same label). The flip flops are clocked synchronously using signal CLK. The outputs of F/F A and F/F B are NANDed together through Gate 1 as are the outputs of F/F C and F/F D through Gate 2. The outputs of Gates 1 and 2 are Invert ORed through Gate 3. Gate 3 provides the signal OUTPUT. The PRESET and CLEAR inputs to all flip flops are set to a logical 1 since they won't be used.

A spreadsheet was set up (Figure 8-2 and Figure 8-3) using an arbitrary CLK period of 12 rows: six low and six high.

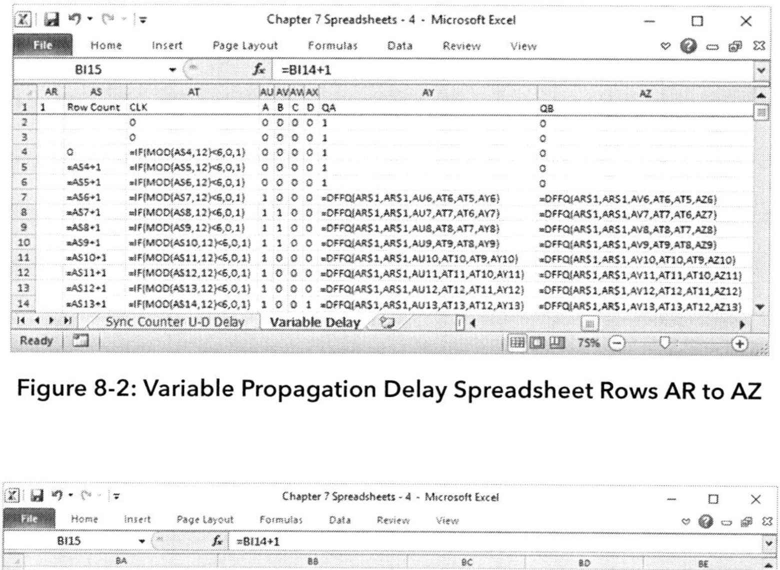

Figure 8-2: Variable Propagation Delay Spreadsheet Rows AR to AZ

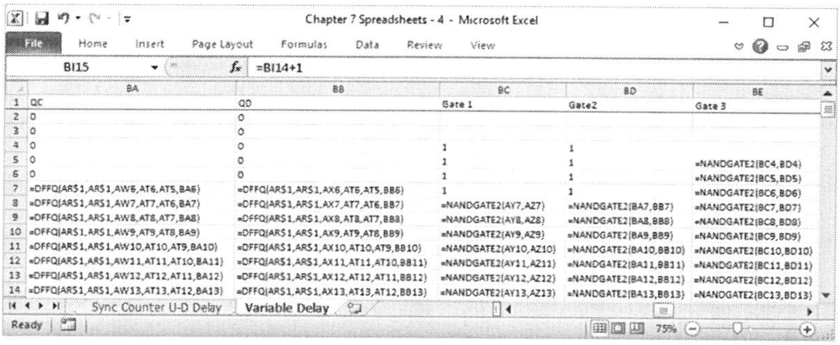

Figure 8-3: Variable Propagation Delay Spreadsheet Rows BA to BE

Chapter 8: *Variable Delay Analysis* | 181

Patterns for signals A, B, C, and D are entered in Columns AU, AV, AW, and AX. Flip flop simulation is performed in Columns AY through BB. It employs the standard timing relationships that we have developed up to now for determining which input rows are used to derive the output. Note that PRESET and CLEAR are connected to Cell AR1 which is simply defined as a 1.

Gates 1, 2, and 3 are defined in Columns BC, BD, and BE. Inputs for a specific output are taken from the previous row (i.e. we implement a one-row delay for these gates).

The data view of the full spreadsheet is given in Figure 8-4. The results are what we would expect from the input pattern.

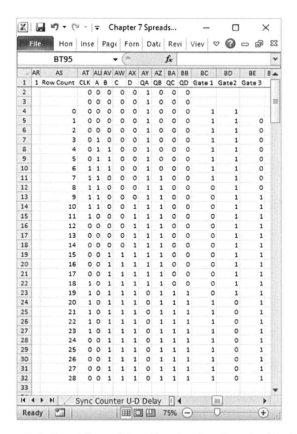

Figure 8-4: Data View of Test Circuit using Standard Delays

Let us now modify the spreadsheet to analyze timing variations.

The first step in doing this is to examine actual delays through each component. For this circuit, we find the following specified for Texas Instruments 74LS00 and 74LS74 devices:

Device	Typical Delay	Maximum Delay
74LS00 (NAND)	9 ns	15 ns
74LS74 (D Flip Flop)	13-25 ns (Clock to Q)	25-40 ns (Clock to Q)

Table 8-1: Timing Specification for the NAND Gate and D Flip Flop

The varying typical and maximum delay times for the 74LS74 depends on whether the output is switching from high to low or low to high. For our analysis, we will simply use the worst case delay time (i.e. 40 ns for the maximum delay) regardless of the switching direction.

Ideally, the spreadsheet would be set up so that the time duration of each row would be a common factor of all the propagation times. However, using actual values, it is very unlikely to find a common factor other than 1. What can be done for this case is to set the spreadsheet row time duration to 5 ns. So the maximum delay for the 74LS00 device would be three rows and the 74LS74 would be eight rows. If the times were not in multiples of 5 ns, the values could just be rounded up to the next multiple of 5 ns. This would check the design for a greater worst case tolerance than is likely to be seen in reality.

For the typical delays, which are not multiples of 5 ns, there are other options as well. Here, we can round the times down. So 9 ns would be taken to be 5 and 13 ns would round to 10. Alternatively, all of the typical values could be rounded up to the next multiple of 5. But if we are analyzing the interaction between typical and worst case delays within a simulation, rounding typical values down and worst case values up usually results in a simulation that stresses the design beyond expected parameters. It is up to the designer to determine if any observed anomalous behavior is an actual design problem. This type of simulation does indicate where the weakest point in the design exists, timing-wise, even though the conditions that brought it about are not likely to occur.

So the values in Table 8-1 are modified to those in Table 8-2.

Device	Typical Delay	Maximum Delay
74LS00 (NAND)	5 ns	15 ns
74LS74 (D Flip Flop)	10 ns (Clock to Q)	40 ns (Clock to Q)

Table 8-2: Simulated times for the NAND Gate and D Flip Flop

Based on the values presented in Table 8-2, the typical timing for the NAND gate equations will take its input values from the preceding row. The flip flops' outputs will update based on the data latched two rows back. Furthermore, let us take the period of the clock to be 100 ns (i.e. a 10 MHz clock). Since the spreadsheet row represents 5 ns, the clock has a period of 20 rows. The formula view of the spreadsheet is given in Figure 8-5 and Figure 8-6.

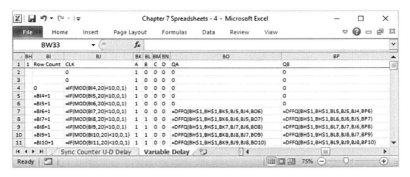

Figure 8-5: Circuit Simulation for Typical Values, Formula View; Columns BH to BP

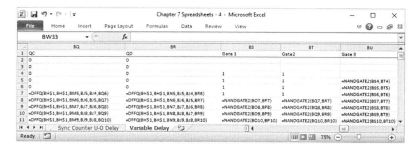

Figure 8-6: Circuit Simulation for Typical Values, Formula View;
Columns BQ to BU

Column BH has the logical 1 reference that is used for the PRESET and CLEAR flip flop inputs.

Column BJ uses a modulo 20 division so that the CLK period is 20 rows.

The inputs for the flip flop D inputs are entered manually in Columns BK to BN. They maintain the same pattern as Figure 8-4.

The Q formulas for the flip flops are in Columns BO to BR. Notice that the formulas look back two and three rows for the CLK transition. That means the transition is sought at the end of the third row back from the preset row or, equivalently, the beginning of the second row back. The simulation begins at Row 7 so the preceding rows just have some initialization values.

Columns BS through BU have the formulas for the three NAND gates. Their inputs are taken from the preceding row, simulating a 5 ns delay.

The data view of the spreadsheet is given in Figure 8-7 and Figure 8-8.

Chapter 8: *Variable Delay Analysis* | 185

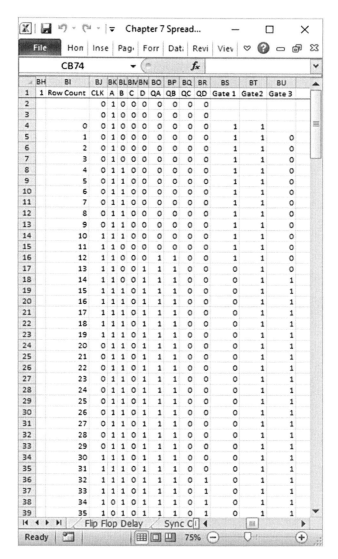

Figure 8-7: Circuit Simulation for Typical Values, Data View; Rows 1 Through 39

186 | Digital Circuit Simulation Using Excel

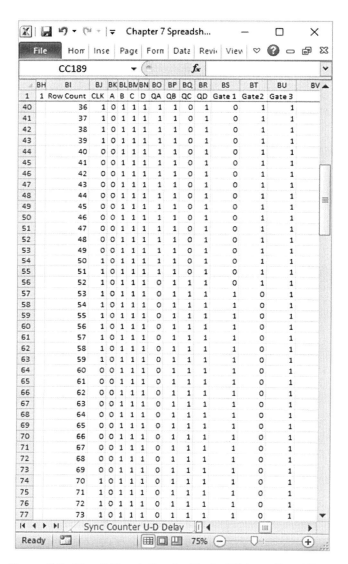

Figure 8-8: Circuit Simulation for Typical Values, Data View; Rows 40 Through 77

Chapter 8: *Variable Delay Analysis* | 187

Looking at the data view, there is nothing remarkable about the results. They look like the results previously obtained in Figure 8-4.

Figure 8-9 shows the timing diagrams that are derived from the simulation for Gates 1, 2, and 3.

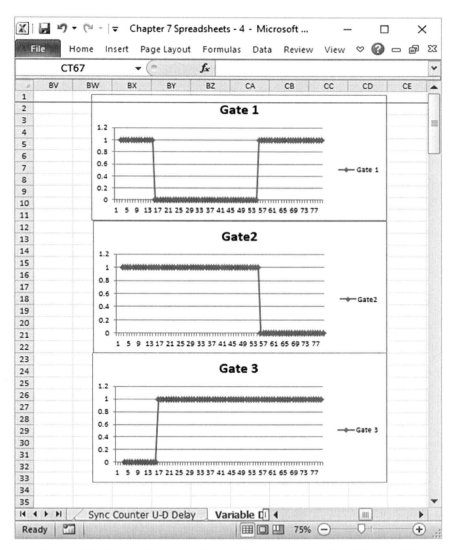

Figure 8-9: Timing Diagrams Derived from Figure 8-7 and Figure 8-8

188 | Digital Circuit Simulation Using Excel

Figure 8-10 and Figure 8-11 show the first few rows for the formula view of the spreadsheet, assuming worst case delays for all devices.

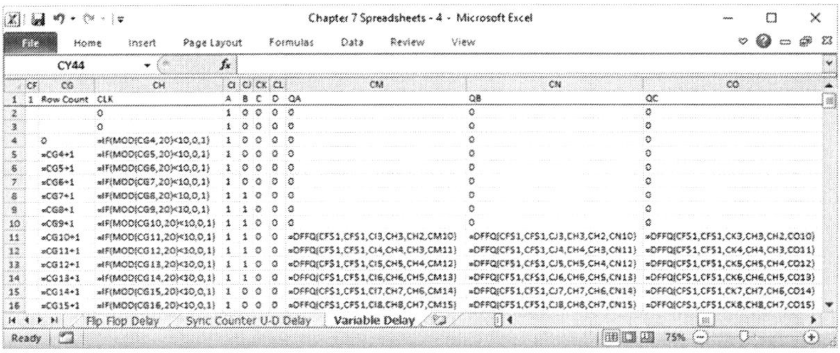

Figure 8-10: Columns CF to CO of the Spreadsheet Formula View using Worst Case Propagation Delays

Figure 8-11: Columns CP to CS of the Spreadsheet Formula View using Worst Case Propagation Delays

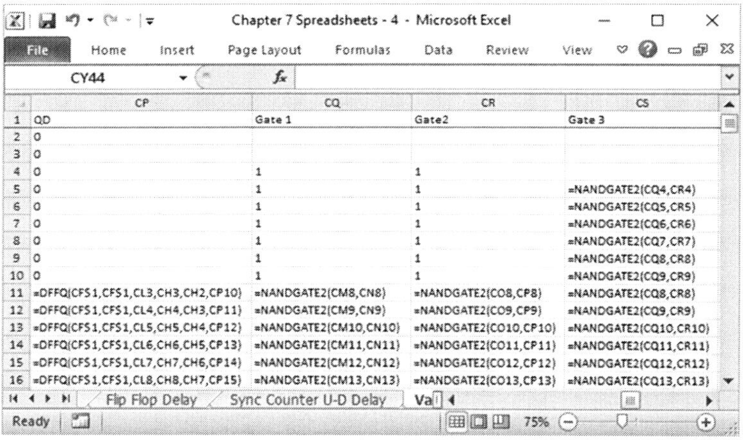

In order to realize the 40 ns propagation delay for the flip flops, the CLK and D inputs used are from 8 rows back. Looking at Cell CM11 in Figure 8-10, the clock (CLK) is examined in Cells CH2 and CH3. The data for the D input comes from Cell CI3.

Notice that the simulation started much later in Row 11. This is because more initialization time was needed since the flip flop formulas referenced input data much earlier in time to simulate more delay.

Figure 8-11 shows the simulation of the three NAND gates in Columns CQ, CR, and CS. For the 15 ns delay, the output formulas had to reference back three rows. So the input cells for the three NAND gates in Row 11 use data from Row 8.

The simulation is run and the timing diagrams for the three NAND gates are shown in Figure 8-12.

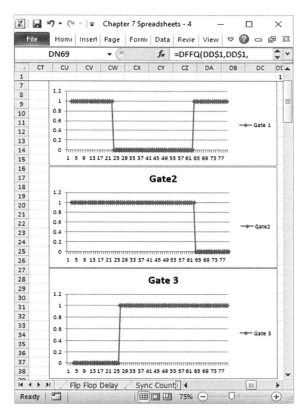

Figure 8-12: Timing Diagrams for the NAND Gates using Maximum Delays for all Devices

The timing diagrams in Figure 8-12 using the maximum delay for all devices look very similar to the diagrams for typical delays (Figure 8-9). The only noticeable difference is that the events in Figure 8-12 are delayed with respect to CLK from those in Figure 8-9. This can be observed by the numerical indices at the bottom of each timing diagram. This is to be expected when maximum delays are used.

Let us consider one additional case. In this instance, we will restore all the devices to their typical values except for those in the path of Flip Flop C. There, the maximum propagation delay values will be used. In Figure 8-13, compare the rows used as inputs for the Flip Flop C formulas (Column DM) to the rows used by inputs by the other flip flops in the same row.

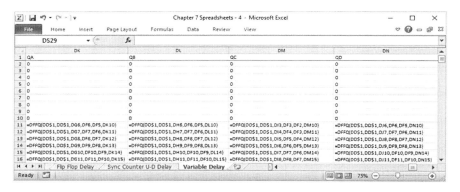

Figure 8-13: Spreadsheet Formula View for Flip Flop Inputs with Maximum Delay in the Path for Flip Flop C

Similarly, Figure 8-14 shows the inputs for the NAND gates. Notice that Gate 2 references input from three rows back (equivalent to 15 ns), while the other gates use inputs from the previous row (5 ns).

Chapter 8: *Variable Delay Analysis* | 191

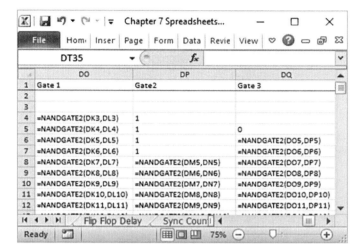

Figure 8-14: Formula View for NAND Gate Inputs with Maximum Delay in the Path for Flip Flop C

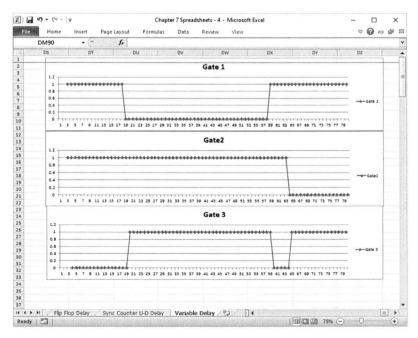

Figure 8-15: Timing Diagrams Showing the Three NAND Gate Signals with Maximum Delay in Flip Flop C's Path

The resulting timing diagrams are shown in Figure 8-15. Notice what happened in this configuration. There is an extra pulse generated by Gate 3 that was not there before.

This situation can be analyzed by looking at the data view of the spreadsheet in this area. See Figure 8-16.

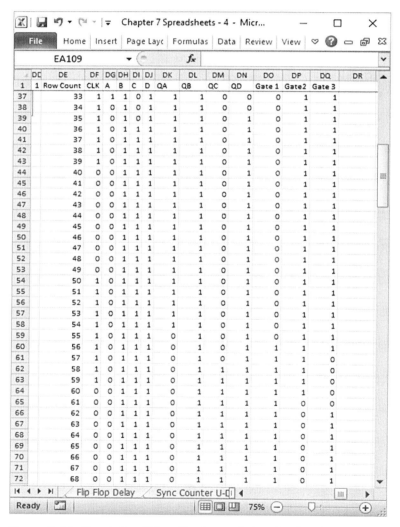

Figure 8-16: Simulation Using Maximum Delays in Flip Flop C's Path

CHAPTER 9

Excel Implementation of Standard 7400 Devices

This section develops Excel simulation functions for some of the standard devices found in the 74xx logic families.

The primary emphasis is on creating generalized functions that can be called from the library. As such, they can be applied to any of the specific 74xx families (e.g. 74LS, 74HC, 74S, etc.). The examples developed in this chapter cite the 74LS parameters. However, they can be customized to any specific family by incorporating the appropriate propagation delays of that family into the corresponding time intervals represented by the rows of the spreadsheet (covered in previous chapters).

Tri-State Devices

Tri-State devices are designed to allow their outputs to be directly connected together. When a device is enabled, it is allowed to drive the connection that its output shares with the outputs of other tri-state devices. The outputs of the devices that are not selected go into a high impedance mode. So only the selected device drives that common point. A device is selected by having its output enabled by a controlling signal.

An example of this device is the 74LS125. It contains four independently controlled tri-state buffers. Each buffer has the following circuit representation:

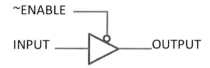

Figure 9-1: Circuit Representation of the Tri-State Buffer

When the ~ENABLE signal is inactive (at a logical 1 level), OUTPUT is in the high impedance state and is effectively isolated from any other devices connected to this point.

When ~ENABLE is active (at a logical 0 level), the signal at OUTPUT equals the signal at INPUT.

Consider an example where the four gates in the 74LS125 device are configured to connect their outputs together so that the signals at each of the four inputs can be independently selected to drive the common output point. Bear in mind that the gates in a single device can be configured independently of each other. That is, their outputs do not have to be connected to each other. We have chosen to configure them as shown in Figure 9-2.

Chapter 9: *Excel Implementation of Standard 7400 Devices* | 197

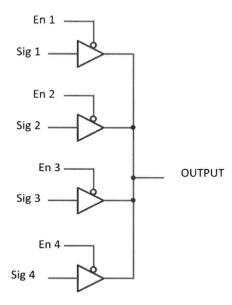

Figure 9-2: Example of Four Tri-State Buffers Driving a Single Point

Each device has one input signal designated with the prefix "Sig" and one Tri-State enable signal using the prefix "En." The spreadsheet implementation is given in Figure 9-3.

	A	B	C	D	E	F	G	H I J	K
1	Sig 1	En 1	Sig 2	En 2	Sig 3	En 3	Sig 4	En 4	3-State
2	0	1	1	1	1	1	1		=IF(B2=0,A2,(IF(D2=0,C2,(IF(F2=0,E2,(IF(H2=0,G2,1))))))
3	0	0	1	1	1	1	1		=IF(B3=0,A3,(IF(D3=0,C3,(IF(F3=0,E3,(IF(H3=0,G3,1))))))
4	1	0	0	1	1	1	1		=IF(B4=0,A4,(IF(D4=0,C4,(IF(F4=0,E4,(IF(H4=0,G4,1))))))
5	1	1	0	1	1	0	1		=IF(B5=0,A5,(IF(D5=0,C5,(IF(F5=0,E5,(IF(H5=0,G5,1))))))
6	1	1	0	0	1	0	1		=IF(B6=0,A6,(IF(D6=0,C6,(IF(F6=0,E6,(IF(H6=0,G6,1))))))
7	1	1	1	0	1	0	1		=IF(B7=0,A7,(IF(D7=0,C7,(IF(F7=0,E7,(IF(H7=0,G7,1))))))
8	1	1	1	1	0	1	1		=IF(B8=0,A8,(IF(D8=0,C8,(IF(F8=0,E8,(IF(H8=0,G8,1))))))

Figure 9-3: Spreadsheet Formula View for Four Tri-State Buffers Driving a Single Point

The first device is defined in Columns A (for Sig 1) and B (for En 1). The second is defined in Columns C and D. The third and the fourth are in Columns E and F and Columns G and H respectively.

The common output node, where all the outputs are tied together, is implemented in Column K. It consists of nested IF/THEN/ELSE statements. It looks at each "En" in turn. If the first "En" signal is high, it checks the next "En." If it finds that there are no active "En" signals, it sets the cell value to 1. If any of the "En" signals are low, the value of the associated "Sig" signal is selected to be the value assigned to the associated cell in Column K and no further searching takes place.

As noted above, if no device is selected (i.e. all of the "En" signals are high), we set the tri-state point to a level of 1. This is a reasonable decision since, in practice, tri-state points are often equipped with pull-up resistors to bring the node to a defined state when no device is selected. This is commonly done for testing purposes to verify the high impedance condition of all outputs by allowing the pull-up resistor to create a high level.

It is interesting to note that the tri-state operation in Column K does not really define a single device. Rather it describes the operation at the common node where the tri-state devices are connected. The four points driving the node can be from four totally separate devices.

A test case is run in Figure 9-4. It is clear that when an "En" is active (low), the tri-state function (Column K) reflects the value of the selected "Sig."

Chapter 9: *Excel Implementation of Standard 7400 Devices* | 199

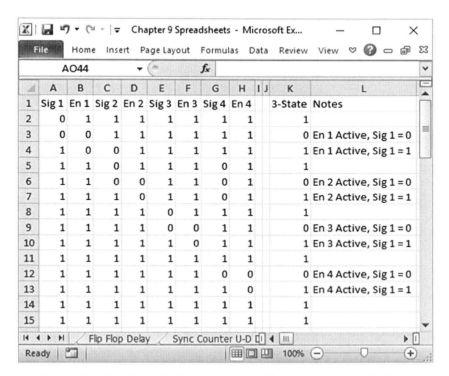

Figure 9-4: Test Case for Four Tri-State Buffers Driving a Single Point

Note that if two "En" signals are active at the same time, this implementation does not detect it or generate an error condition. **(Can you see a way of doing this? A simple approach is to create an independent spreadsheet column solely for the purpose of detecting this situation and generating an error notification).**

A Visual Basic library function can be created for tri-state devices. This would be the equivalent to Column K in Figure 9-3 and

Figure 9-4. It will be defined to act as a node representing the connection of up to four tri-state devices. The function is:

```
Function TriState(EN1, SIG1, EN2, SIG2, EN3, SIG3, EN4, SIG4)
    If EN1 = 0 Then
    TriState = SIG1
    Else
        If EN2 = 0 Then
        TriState = SIG2
        Else
            If EN3 = 0 Then
            TriState = SIG3
            Else
                If EN4 = 0 Then
                TriState = SIG4
                Else
                    TriState = 1
                End If
            End If
        End If
    End If
End Function
```

If the common node consists of fewer than 4 devices connected together, the unused "En" signals can be assigned to a cell with a constant value of 1. This is shown in Figure 9-5, where the formula view of the spreadsheet shows three devices connected at the common output point. Passed parameters EN4 and SIG4 are connected to Cell A30 which has the value of 1. Functionally, it is only necessary to assign EN4 to a value of 1, but since it is required to assign a value to all passed parameters, we selected the value of 1 for SIG4 as well.

Chapter 9: *Excel Implementation of Standard 7400 Devices* | 201

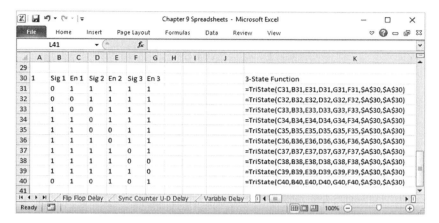

Figure 9-5: Formula View for a Function Call Connecting Three Tri-State Devices

The data view of the functions is shown in Figure 9-6 with comments for the test given in Column L.

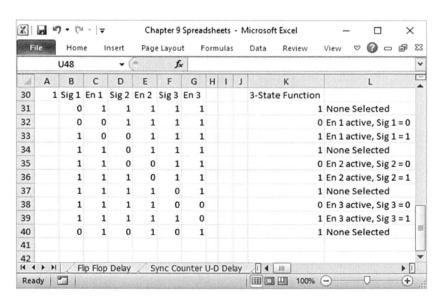

Figure 9-6: Data View for a Function Call Connecting Three Tri-State Devices

Typical data sheets indicate that the specified propagation delays are approximately as follows for the 74LS125:

Parameter	Typical	Maximum
Propagation Delay Input to Output	7 ns	18 ns
Propagation Delay Enable to Output	12 ns	25 ns

In the previous chapter, we explained how the different combinations of delay times can be simulated in the spreadsheets by finding common factors. We will not revisit the subtleties of that topic in this chapter, but rather will focus on simulation considerations using broad simplifying assumptions. In this case, let us just assume that the "Enable to Output" propagation delay is twice that of the "Input to Output" delay. We can then let the "Input to Output" delay equal one row making the "Enable to Output" delay equal to two rows.

Let us take the test that was shown in Figure 9-5 and Figure 9-6 and expand it to account for propagation delays. Each test case that was originally one line is expanded to 6 lines, creating an input sequence in time. The first few rows of the formula view of the spreadsheet are given in Figure 9-7.

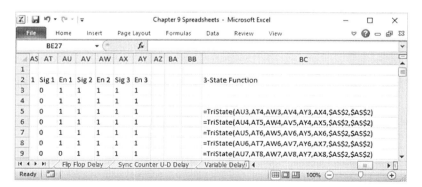

Figure 9-7: Formula View of the Spreadsheet Simulating Tri-State Devices with Propagation Delays

Chapter 9: *Excel Implementation of Standard 7400 Devices* | 203

Essentially, the value for the "En" signals is taken from two rows back and the value for the "Sig" signals is taken from one row back.

The data view of the spreadsheet (Rows 1 to 68) is given in Figure 9-8.

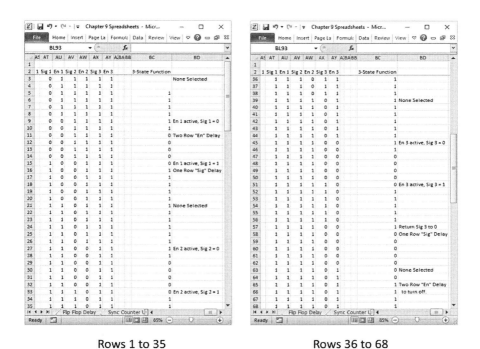

Rows 1 to 35   Rows 36 to 68

Figure 9-8: Data View of the Spreadsheet Simulating Tri-State Devices with Propagation Delays

Notice that in Row 9 the En 1 signal becomes active. There is a two-row delay (to Row 11) before Sig1's 0 value is seen at the tri-state point. When Sig 1 changes from 0 to 1 in Row 15, there is a one-row delay (to Row 16) before the change is propagated to the output. Each of the "En" signals is activated in this simulation and the device's response can be verified. Finally, in Row 63, all of the "En" signals are disabled. We can see a two-row delay to Row 65 before the output goes into the tri-state condition.

One other consideration concerns how we can implement a tri-state node when there are more than four devices that are driving the common node. One way is to define additional Excel functions having more than the four devices corresponding to the 74LS125. There is another way to do this without creating additional functions. Consider connecting 8 tri-state device outputs. A circuit is shown in Figure 9-9.

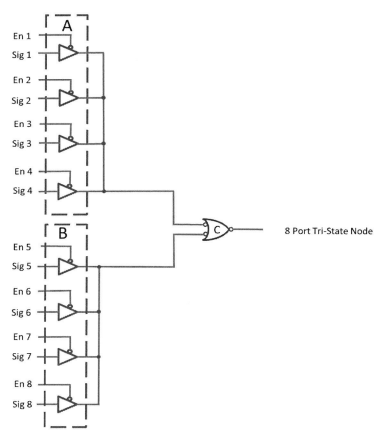

Figure 9-9: Circuit Implementation for an Eight-Port Tri-State Node

The first four signals, Sig 1 to Sig 4, and their respective enable signals, En 1 to En 4, are handled by the quad tri-state device labeled

"A" in the schematic. Similarly, the second set of four devices is realized using device "B." Recall that only one of these eight ports will be active at a time and we have defined the output to these devices as being in a logical high (1) when no devices are selected. Therefore, the simplest way to combine them is with an OR function whose output goes to a low when either of its inputs is a low. So if no tri-state device is selected or if the selected port is in a 1 state, the "8-Port Tri-State Node" in Figure 9-9 will be 1. If any selected port is 0, the output is a 0. This implementation uses an "Invert NOR" function which is functionally equivalent to an AND (remember DeMorgan's Theorem).

The Invert NOR gate should be set up with a zero propagation delay. That is, the output of the gate in the spreadsheet should be derived from the inputs in the same row. The reason is that this is a virtual device that is simply used to logically connect the eight ports together. It is not a physical gate. Also, if we are going to simulate different propagation delays through the tri-state devices, the delays should be associated with the tri-state buffers and not distributed through the virtual Invert-Nor gate that is common to all the buffers.

Let us see how this works in a spreadsheet. Consider Figure 9-10 and Figure 9-11, which have the first few rows of the formula view for this simulation.

Figure 9-10: Eight-Signal Tri-State Port Spreadsheet Formula View Columns P through AI

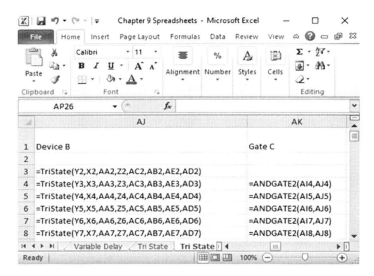

Figure 9-11: Eight-Signal Tri-State Port Spreadsheet Formula View Columns AJ and AK

The test case is set up demonstrating buffers being enabled from both Device A and Device B. Device A has a Sig to Output delay of one row and an En to Output delay of two rows as before. Just to have the devices operate with different parameters, Device B operates with a one-row delay for both Sig and En.

The data for the "Sig" and "En" signals are entered manually to provide a test case. Columns AI and AJ define the two four-port tri-state devices. Gate C in Column AK combines the two four-port nodes into a single eight-port node.

Figure 9-12 shows the execution of the test case. No device is selected until Row 5. In Row 5, En 1 is enabled, allowing Sig 1 to pass through. At this time Sig 1 is 0. In Row 7, we see the tri-state point going to 0, which means that the enabled device is now driving the common output connection. Notice that only one 0 was passed through to the output since the tri-state point returns to 1 in the next row, Row 8. The one zero is due to the fact that the En signal requires

Chapter 9: *Excel Implementation of Standard 7400 Devices* | 207

a two-row delay. So the 0 on Sig 1 in Row 5 did not pass through; only the 0 in Row 6 did. For the rest of the test where this device is enabled, there is just a one-row delay seen from Sig to the output. En 1 is disabled in Row 11; however, the buffer continues to pass data until Row 13 representing the two-row delay from En to Output.

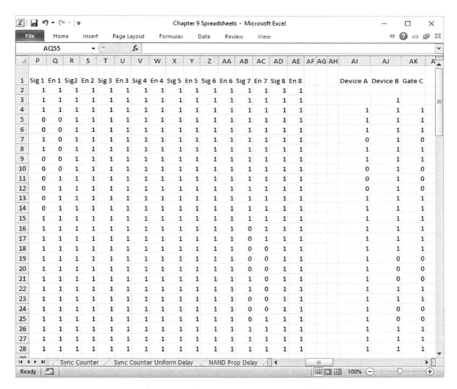

Figure 9-12: Eight-Signal Tri-State Port Spreadsheet Data View

A second test for Sig 7 and En 7 begins at Row 18. En 7 is activated in Row 18. The input, Sig 7 is a 0, but the delay causes the 0 to be propagated in Row 19. Here, there is a one-row delay for the En signal. Since the Sig to Output delay is also one row, the 0 seen in Row 19 is the value of Sig 7 in Row 18. The Sig 7 input stays in the 0 state until Row 20 so the three input 0s (Row 18 to 20) are seen at the tri-state point in Rows 19 to 21. Sig 7 goes to a 1 in Rows 21 and 22 which are seen at the output in Rows 22 and 23. En 7 is disabled in Row 25 and the output is disabled in Row 26.

Decoders

Decoders take a binary-encoded number and activate an output associated with that number. An illustrative example is the 74LS43 one of ten decoder. It takes a four-bit binary input number and activates one of 10 output signals. A truth table for the device is shown in Table 9-1. Notice that the active state at the output is a zero. O0 through O9 will only be low when the binary number coded by ABCD is equal to its subscript.

Inputs				Outputs					
A	B	C	D	O0	O1	O2	...	O8	O9
0	0	0	0	0	1	1		1	1
0	0	0	1	1	0	1		1	1
0	0	1	0	1	1	0		1	1
.									
.									
.									
1	0	0	0	1	1	1		0	1
1	0	0	1	1	1	1		1	0
1	0	1	0	1	1	1		1	1
.									
.									
.									
1	1	1	1	1	1	1		1	1

Table 9-1: Truth Table for the One of Ten Decoder

When ABCD encodes a number from 10 (1010) to 15 (1111), all outputs remain in the high state.

There are two ways to implement this as a function. One approach is to create ten separate functions in the library, one representing each output. A second way is to only create one function but pass the target decoded number to it as a parameter. We will take the latter approach.

The function will have five passed parameters. Parameters A, B, C, and D will represent the encoded binary number as shown in Table 9-1. The encoded binary number is expressed as:

$$\text{Encoded Number} = A^8 + B^4 + C^2 + D$$

The fifth parameter is called "Value." It is the value of ABCD which will cause the function to return a 0.

Given these parameters, the decoder function is defined as:

```
Function Decoder10 (A, B, C, D, Value)
    If 8 * A + 4 * B + 2 * C + D = Value Then
    Decoder10 = 0
    Else
        Decoder10 = 1
    End If
End Function
```

The function was called "Decoder10" because it decodes one of ten values. If we were to implement a one of eight decoder (e.g. 74LS138), it would be called "Decoder8."

The formula view for the first few rows of a test spreadsheet is shown in Figure 9-13.

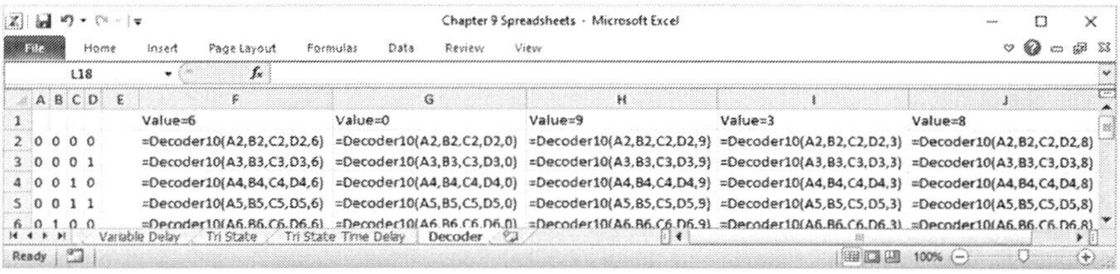

Figure 9-13: Formula View of a Spreadsheet to Test a 1 of 10 Decoder

Columns A, B, C, and D contain a binary progression which will represent the counts 0 through 15 where A is the most significant bit and D is the least significant bit.

Five tests are run where the target decoded value is shown in Row 1. The value can be seen as the fifth passed parameter in the function calls in the respective columns.

The data view of the spreadsheet is shown in Figure 9-14. It can be seen that the value of 0 in each test (Columns F through J) occurs where the encoded value is equal to the target value passed to the function.

Chapter 9: *Excel Implementation of Standard 7400 Devices* | 211

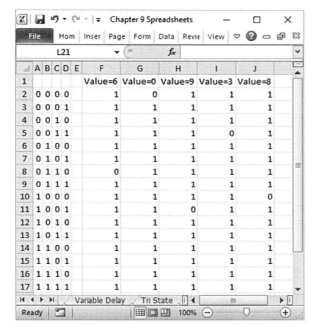

Figure 9-14: Data View of a Spreadsheet to Test a 1 of 10 Decoder

The timing information for this device states that the typical propagation time may be either 15 or 20 ns depending on whether the internal circuitry of the device needs to propagate through either two or three levels of logic. Similarly, worst case propagation is specified as 25 or 30 ns which, again, is dependent on whether there are two or three levels of logic to decode. A prudent designer would normally check for all extreme cases. For this illustrative example, a one-row propagation delay from input to output will be used. Figure 9-15 shows the formula view of the resulting spreadsheet.

212 | Digital Circuit Simulation Using Excel

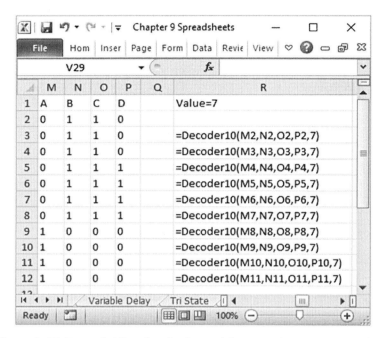

Figure 9-15: Formula View for the Spreadsheet Implementing a 1 of 10 Decoder with a One-Row Propagation Delay

The inputs to the device are encoded in Columns M through P representing the binary count ABCD. They are the first four passed parameters and are selected from the previous row. If we define this particular implementation to decode the number 7, 7 becomes the last passed parameter.

The data view of the spreadsheet is given in Figure 9-16. Row 5 shows the encoded 7 appearing at the input. The device responds by outputting a 0 in Row 6. Then the input changes to an encoded 8 in Row 9 causing the output to be deactivated in Row 10.

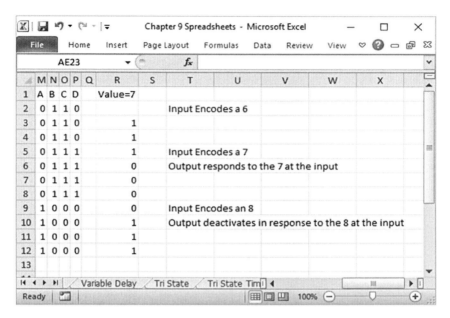

Figure 9-16: Data View for the Spreadsheet Implementing a 1 of 10 Decoder with a One-Row Propagation Delay

One Shot (Monostable Multivibrator)

The one shot is a circuit that generates a single pulse triggered by some condition appearing on its inputs. The duration of the pulse is often determined by a time constant set by analog components (resistors and capacitors). Device data sheets define what component values to use in order to obtain a specific output pulse duration.

214 | Digital Circuit Simulation Using Excel

The function of an idealized, positive edge triggered one shot is shown in Figure 9-17.

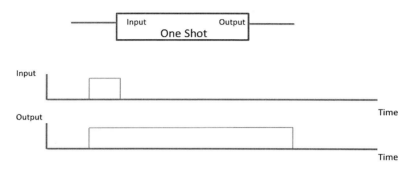

Figure 9-17: Basic One Shot Functionality

The external resistor and capacitor used to set the time constant for the output pulse duration is not shown.

The output is initially at 0. Then a triggering pulse is presented at the input. Being positive edge triggered, the rising edge of the input pulse starts the output pulse. The output pulse will continue (here shown in an active high state) until the time constant defined by the resistor and capacitor has elapsed.

In many cases, one shot devices are retriggerable. That is, if there is another rising edge on the triggering input before the output has timed out, the output pulse is extended for another time constant beginning with the rising edge of the second input pulse. This is shown in Figure 9-18.

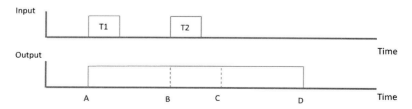

Figure 9-18: Retriggering a One Shot

The original triggering pulse, T1, initiates the output pulse at Point A. The normal non-retriggered output would end at Point C. However, before Point C is reached, a second triggering pulse occurs at Point B. This has the effect of restarting the output pulse duration timer. The result is that the output now terminates at Point D where the duration from B to D is the normal timeout period. The resulting pulse extends from Point A to Point D. Of course if another retrigger occurred before Point D, the output pulse would have been extended even further.

The strategy for implementing the one shot function in Excel is to initialize a count when the triggering condition is detected. After that (assuming for a moment that retriggering does not occur), the count is decremented in each subsequent spreadsheet row. The count reaching zero indicates that the one shot timer has elapsed.

It is evident that the initial count is found by dividing the desired output pulse duration by the time period that each spreadsheet row represents.

If a retriggering event takes place, the count is simply restarted from that point.

This function provides a different type of output than the others that have been described up to now. The output (or returned value) is not binary (a 1 or 0). Rather it is a count generating all integers from the initial count down to 0. In order for this to work with other digital circuit components, we can allocate another spreadsheet column to convert the count to a digital signal. This is a simple conversion. Essentially, the conversion looks at the count. If it is 0, it generates a 0. If it is not 0, it provides a 1. To illustrate, let us assume that the one shot is implemented in Column A of the spreadsheet and the conversion function is in Column B. The conversion entry in Cell B8 would be:

=IF(A8=0,0,1)

Subsequent rows are copied down.

Note that the conversion can be modified to give an inverted pulse if so desired. To do this, the "then/else" actions in the IF statement are reversed as follows:

$$=IF(A8=0,1,0)$$

Using the 74LS123 as a model. This device has:

- three inputs used to control the triggering of the output pulse
- connection points for an external resistor and capacitor that define the duration of the output pulse
- an inverted and non-inverted output

Manufacturer's data sheets explain how to select the resistor and capacitor to provide the desired output pulse width and will not be covered here.

The three triggering inputs are normally designated as A, B, and CLEAR. The CLEAR input, in addition to being able to trigger the output, also provides the capability of terminating an existing output pulse. The operating conditions are shown in Table 9-2. Note that there are two outputs for the device that are not explicitly shown in the Table. One provides an active high output pulse and the other an active low pulse. In the table, we just note the conditions that do and do not trigger an output.

INPUTS			OUTPUT RESPONSE
CLEAR	A	B	
L*	X	X	No Trigger
X	X	L	No Trigger
X	H	X	No Trigger
H	↓	H	Trigger
H	L	↑	Trigger
↑	L	H	Trigger

Table 9-2: Triggering Conditions for the 74LS123

Notes for Table 9-2:

- * When CLEAR = 0 any existing output pulse in terminated
- X = Don't Care
- H = High, or 1
- L = Low, or 0
- ↓ = Negative edge
- ↑ = Positive edge

Given the behavioral model described above, the following Visual Basic Function is derived.

```
Function ONESHOT (A, AX, B, BX, CL, CLX, COUNT, INITCOUNT)
   If CL = 0 Then
   ONESHOT = 0
   Else
      If A = 0 And AX = 1 And B = 1 And BX = 1 And CLX = 1 Then
      ONESHOT = INITCOUNT
      Else
         If B = 1 And BX = 0 And A = 0 And AX = 0 And CLX = 1 Then
         ONESHOT = INITCOUNT
         Else
            If CLX = 0 And A = 0 And AX = 0 And B = 1 And BX = 1 Then
            ONESHOT = INITCOUNT
            Else
               If COUNT = 0 Then
               ONESHOT = 0
               Else
                  ONESHOT = COUNT - 1
               End If
            End If
         End If
      End If
   End If
End Function
```

The passed parameters are defined as:

    A: The cell specified for the signal on the A input

    AX: The cell in the row preceding the one specified for "A" above

    B: The cell specified for the signal on the B input

    BX: The cell in the row preceding the one specified for "B" above

    CL: The cell specified for the signal on the CL input

    CLX: The cell in the row preceding the one specified for "CL" above

    COUNT: The value provided by the One Shot function in the previous row

    INITCOUNT: The value provided by the One Shot function when initially triggered

Let us analyze the function to see how the decisions are made.

First the Clear (CL) signal is checked to see if it is in an active 0 state. If so, the value of the function is set to 0 and then terminates.

If CL is not active low, the function begins to look for other triggering conditions.

First it checks for a negative transition on A. It does this by seeing if the parameter A (the present value) is 0 and AX (the previous value) is 1. Also inputs CL and B must also be checked to see if they are in the correct states to allow A to successfully trigger an output pulse. The data sheets do not specify for how long CL and B must remain in the correct states before or after the transition on A for the triggering to occur. For our simulation, we will require that they are in their valid states, to allow A to generate a trigger, in both the previous and current spreadsheet rows. That is, they must be in the correct states

Chapter 9: *Excel Implementation of Standard 7400 Devices* | 219

in the intervals (rows) immediately before and immediately after the transition. So B and BX are checked to see if they are both 1. CLX is checked to see if it is also 1. CL does not need to be checked here since the first test in the function verified that CL is 1. If the triggering conditions were met, then the function returns the value INITCOUNT.

If the previous check did not generate an output pulse, the function checks to see if a positive transition occurs on Input B while A=0, AX=0, and CLX=1. If this condition is detected, the function returns INITCOUNT.

The next check looks for a positive transition on CL while A=0, AX=0, B=1, and BX=1. Since it was established that CL=1, seeing if CLX=0 along with the necessary conditions for A and B would verify that the triggering condition was met and INITCOUNT would be returned.

If none of the triggering conditions were met, then the count is unchanged if it was previously 0 or decremented if it was not previously 0.

A spreadsheet for testing the operation of the one shot is given in Figure 9-19.

	L	M	N	O	P	Q	R
1	Count Value	Clear	A	B	Q		Digital Q
2	7	1	1	1	0		=IF(Q2=0,0,1)
3		1	1	1	0		=IF(Q3=0,0,1)
4		1	1	1	=ONESHOT(N3,N2,O3,O2,M3,M2,Q3,L2)		=IF(Q4=0,0,1)
5		1	1	1	=ONESHOT(N4,N3,O4,O3,M4,M3,Q4,L2)		=IF(Q5=0,0,1)
6		1	1	1	=ONESHOT(N5,N4,O5,O4,M5,M4,Q5,L2)		=IF(Q6=0,0,1)
7		1	0	1	=ONESHOT(N6,N5,O6,O5,M6,M5,Q6,L2)		=IF(Q7=0,0,1)
8		1	0	1	=ONESHOT(N7,N6,O7,O6,M7,M6,Q7,L2)		=IF(Q8=0,0,1)

Figure 9-19: Formula View of the One Shot Test Spreadsheet

The three inputs, CLEAR, A, and B are assigned Columns M, N, and O. The one shot function is in Column Q and the conversion of the one shot count to a digital signal is in Column R.

We have also specified that the INITCOUNT parameter value is defined in Cell L2. It could also have been written into the parameter list where the passed parameters are called out in Column Q. However, by defining it in this one cell, the value can be easily changed in this one location and save the step of copying down Column Q. In this case, we have set the count to 7.

To demonstrate, we have simply assumed that the propagation delay for each input parameter is one row. As with the other examples discussed up to now, these can be set to any delay value.

The initial rows for this test are given in Figure 9-20. Comments indicating relevant events are entered in Column T.

The first event occurs in Row 7 where there is a trigger arising from the negative transition for Input A. This is followed by the initial count of 7 being output by the one shot function in Row 8. Without any further triggering, the count decrements down to 0 in Row 15. Note that Column R shows a pulse being generated during the non-zero values provided by Column Q

Row 23 shows another pulse being generated by the triggering on Input B. The count initializes to 7 in Row 24 and begins to decrement. However, this time there is a retriggering event by Input B in Row 27 which restarts the counter to the initial count. Decrementing continues until Row 31 where there is one more retriggering event generated by Input A. This time there is no further retriggering and the count reaches 0 in Row 39 and the output pulse terminates.

Chapter 9: *Excel Implementation of Standard 7400 Devices* | 221

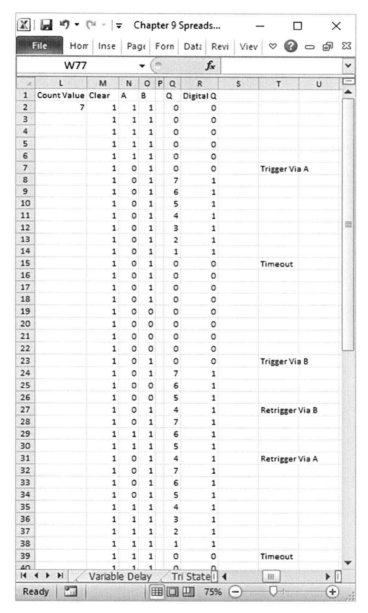

Figure 9-20: Data View of Rows 1 through 39 of
the One Shot Test Spreadsheet

222 | Digital Circuit Simulation Using Excel

Figure 9-21 gives the rest of the test spreadsheet while Figure 9-22 shows the full test represented in timing diagrams where each input, the one shot count, and digital (or binary) output is shown.

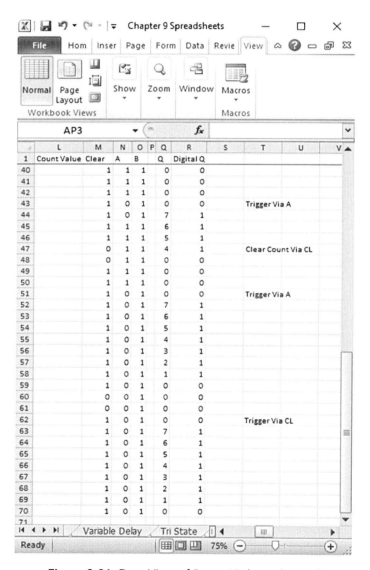

Figure 9-21: Data View of Rows 40 through 70 of the One Shot Test Spreadsheet

Chapter 9: *Excel Implementation of Standard 7400 Devices* | 223

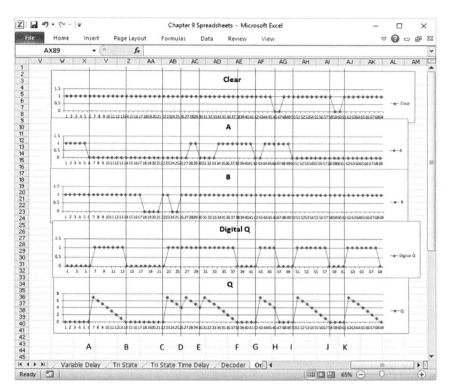

Figure 9-22: Timing Diagrams for the One Shot Test

In the timing diagrams, the major events are indicated with letters at the bottom of the diagram just below the "Q" time line. Vertical lines are included so that concurrent events can be easily found. The events are described in the following. Note: Events A through F were already discussed as part of the description for Figure 9-20.

Event A: The one shot is initially triggered using the falling edge of Input A. CLEAR and B are both equal to 1 so the triggering condition is met. Notice on the next datapoint after Event A, the count (Graph Q) initializes to 7 and the digital representation of Q ("Digital Q") goes to 1.

Event B: The count is observed decrementing to 0 and the Digital Q signal goes back to zero. The one shot has timed out.

Event C: Triggering occurs on the rising edge of B while CLEAR=1 and A=0. Q gets set to 7 and the decrementing begins.

Event D: The count has not decremented to 0. The count gets to 4 and the one shot is retriggered by another rising edge of Input B. The next datapoint shows the counter being set again to 7.

Event E: Again there is triggering before the count can be fully decremented. This time the triggering was the result of a falling edge on Input A. Notice that CLEAR and B are in the correct states to set up a retriggering by A

Event F: The count finally decrements to 0. Notice how retriggering extends the output pulse on the signal Digital Q. Compare the width of the Digital Q pulse between Events A and B with the pulse between C and F

Event G: Another pulse is originated. This time it is the result of a falling edge on Input A.

Event H: By the CLEAR signal going to 0, the count is aborted when it got down to 4 and on the next datapoint it is set to 0. A short output pulse is seen on the Digital Q signal.

Event I: Another trigger is set by Input A

Event J: The count goes to zero producing the normal, non-extended output pulse on the Digital Q signal.

Event K: Here the CLEAR signal is used to trigger another pulse. The conditions are met since A=0 and B=1 and there is a rising edge on CLEAR.

References [1] and [9] catalog generic functionality for the 74XX devices.

Reference [9] compares the timing and electrical characteristics of the common logic families. It also describes the structure of tri-state outputs. Further logic family comparisons can be found in Reference [4].

Additional information on one shot functionality can be found in Reference [8].

Bibliography

[1] Agarwal, C. B. *Digital Principles and Circuits*. Mumbai: Himalaya Publishing House, 2006.

[2] Ashenden, Peter J. *Digital Design: An Embedded Systems Approach Using Verilog*. Burlington, Massachusetts: Morgan Kaufmann Publishers, 2008.

[3] Ferdjallah, Mohammed. *Introduction to Digital Systems Modeling, Synthesis, and Simulation using VHDL*. Hoboken, New Jersey: Wiley, 2011.

[4] Harris, David, and Sarah Harris. *Digital Design and Computer Architecture: From Gates to Processors*. San Franscisco: Morgan Kaufmann Publishers, 2007.

[5] Holdsworth, Brian, and Clive Woods. *Digital Logic Design*. Oxford: Newnes, 2002.

[6] Karris, Steven. *Digital Circuit Analysis and Design with Simulink Modeling and Introduction to CLPDs and FPGAs*. Freemont, California: Orchard Publications, 2007.

[7] Mazzurco, Anthony. *An Introduction to Numerical Analysis Using Excel*. Denver: Outskirts Press, 2019.

[8] Peatman, John B. *The Design of Digital Systems*. New York: McGraw-Hill, 1972.

[9] Singh, A. K., Manish Tiwari, Arun Prakash. *Digital Principles Switching Theory*. New Delhi, New Age International Ltd, 2007.

Index

A
AND Gate 6, 9, 17-21
Associative Rule 27
Asynchronous Decade Counter 91-109, 126, 153-164

C
Circular Reference 57

D
Decoder 208-213
DeMorgan's Theorem 9
DeMorgan's Theorem 37, 205
Don't Care 217
Don't Care 36
Dynamic Simulation 38, 57, 140

E
Excel Function
    AND 17, 20, 21, 63, 65
    IF, Then, Else 16, 19, 30, 63, 65, 198, 215
    OR 23, 24
EXCLUSIVE-NOR Gate 12, 32-33
EXCLUSIVE-OR Gate 11, 25-29

F
Flip Flops
    D Type 61-80, 111, 118, 150-153, 180, 182
    J-K Type 81-89, 92-93, 99

H
High Impedance 195, 198
Hold Time 62

I
Inverter 5, 9, 15, 37, 41, 48, 100, 141
INVERT-OR Gate 10, 37, 41, 119

K
Karnaugh Maps 44, 112-114

M
Macro-Enabled Workbook 41
Modular Arithmetic 18, 19, 117, 150, 184

N
NAND Gate 8, 21, 37, 42, 45, 51, 61, 100, 115, 133-139, 141, 182, 183, 189
NOR Gate 12, 29-31

O
One Shot 213-224
OR Gate 7, 23-25

P
Parity 11, 27, 29
Passed Parameters 41, 77, 78, 200, 209, 210, 212, 218, 220
Propagation Delay 38-41, 55-61, 68-73, 133-170, 179-193, 202, 211-213
Pull-Up Resistor 68, 97, 99, 198

Q
Quine-McCluskey Method 44

R
Race Condition 38, 57, 61, 140
Retrigger 214-215, 220-224
R-S Latch 51-61, 94, 101, 108

S
Setup Time 62, 69
Shift Register 67-72
Spurious Signal 148, 193
Static Simulation 38, 57, 140, 144
Sum of Products 44
Switching Hazard 57, 61, 140, 148
Synchronous Binary Counter
 111-131, 164-178

T
Toggle Mode 81, 84, 94, 99
Transient Count 106-109, 149,
 156-160, 162-164
Transient State 57
Tri-State Devices 195-208

V
Visual Basic 35, 41-43, 45, 77-79,
 82-83, 199, 217